# THE WELFARE OF FUTURE CHILDREN

# THE WELFARE OF FUTURE CHILDREN

## REPRODUCTIVE ETHICS AND DISABILITY SCREENING

*Rebecca Bennett*

BLOOMSBURY ACADEMIC

LONDON · NEW YORK · OXFORD · NEW DELHI · SYDNEY

BLOOMSBURY ACADEMIC
Bloomsbury Publishing Plc
50 Bedford Square, London, WC1B 3DP, UK
1385 Broadway, New York, NY 10018, USA
29 Earlsfort Terrace, Dublin 2, Ireland

BLOOMSBURY, BLOOMSBURY ACADEMIC and the Diana logo are trademarks
of Bloomsbury Publishing Plc

First published in Great Britain 2025



A catalogue record for this book is available from the British Library.

A catalog record for this book is available from the Library of Congress.

ISBN: PB: 978-1-3503-4435-8
HB: 978-1-3503-4434-1
ePDF: 978-1-3503-4437-2
eBook: 978-1-3503-4436-5

Typeset by Newgen KnowledgeWorks Pvt. Ltd., Chennai, India
Printed and bound in Great Britain

To find out more about our authors and books visit www.bloomsbury.com
and sign up for our newsletters.

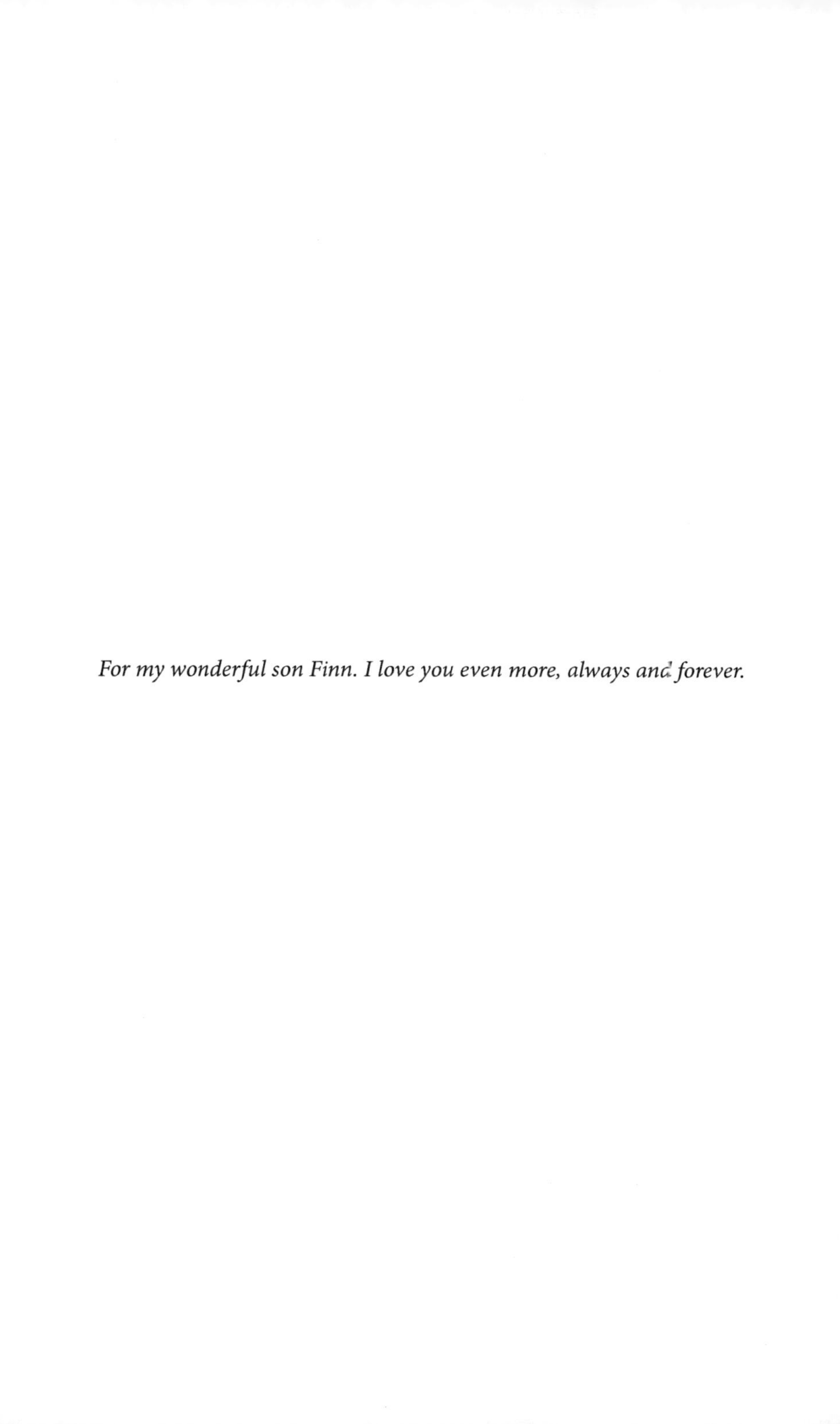

*For my wonderful son Finn. I love you even more, always and forever.*

# CONTENTS

# ACKNOWLEDGEMENTS

The first draft of this book was written during a six-month sabbatical from work, at my dining room table watching the squirrels, cats and birds enjoying my garden while I sat watching and typing, often thinking this book would never be finished. I have a huge thanks to say to my son Finn and my partner Rob (aka Bob). They loved me, supported me, listened patiently to me when I lost faith in the project, encouraged me, brought me hundreds of cups of tea and coffee and generally did what they always do – make my life the wonderful experience that it is. I am a very lucky person to have both of them in my life.

This book is dedicated to my son. I quite selfishly inflicted life on him, but his existence is by far the single most wonderful thing to ever happen to me. He is my life, my joy and my inspiration. He is the bravest, brightest and most brilliant human being I have ever encountered, and I am so incredibly proud to be his mum.

In the thirty years I have worked at the Centre for Social Ethics and Policy (CSEP) at the University of Manchester, I have worked with some truly inspirational people and developed some really treasured friendships.

This journey started with being interviewed by John Harris and being employed to work closely with him on a research project. Working with John was a true privilege, and while I don't agree with a lot of what he argues, disagreeing with him has been a life-changing experience and inspired much of my work in this area, including this book.

Also at that interview thirty years ago were Margot Brazier and Charles Erin. I am ashamed to say that at the time of the interview I had never heard of Margot and the massive role she has played in medical law in the UK and internationally. Margot is of course a brilliant lawyer and law historian, but more than that, she is an inspiration and role model for a huge number of people who have either had the luck to work with her or be taught or supervised by her.

Charles remains one of my close friends. We had many adventures together during our time working together and organizing conferences all over Europe. Together with Valerie Bernard and Betty McGrath, we got to visit some amazing places, meet and work with some brilliant people and get some surely award-winning hangovers! Some of those amazing people include Inez de Beaufort, Matti Häyry, Tuija Takala, Lucy Frith, Heather Draper, Bobbie Farsides, Anton Vedder, Heta Gylling, Mark Cutter, Maurizio Mori, Alan Cribb, Emma Cave, Sara Fovargue, Suzanne Ost, Sheelagh McGuiness, John Coggan, Sue Nelson, David Lawrence, Richard Huxtable, Simo Vehmas, Mike Parker, Suzanne Van de Vathorst, Amel Algrahni, Danielle Griffiths, David Lawrence, Jonathan Glover and Medard Hilhorst.

A special mention goes to Søren Holm. One of the first jobs I was asked to do when starting at the University of Manchester was to phone Søren in Denmark. I am sure he wondered who this gabbling woman was, but in our thirty years of friendship, he has been a loyal and wonderful friend and colleague and now an asset to the University of Manchester.

This book would not have been written without the inspiration and support provided by my colleagues at CSEP. Sarah Devaney and Alex Mullock, currently co-heads of our centre, lead with humour and passion and are the perfect partnership to continue the wonderful leadership of our centre that was started by John Harris and Margot Brazier. The many students I have taught since 1993 have all influenced my research and my thinking about these issues. This includes the many students who have become colleagues, collaborators and friends, including Aeron Haworth, James Anderson, Nicola Williams, Costas Kanaris, Dunja Begovic, Chloe Romanis, Craig Purshouse, Jenny Kruzinna, Mark Josef Rapa, Sacha Waxman, Daniel Bianchi, Patrick Allen, Patrick Quinn, Maria de Jesus Medina Arellano, Ed Horowitz, Ioannis Ntanos, Kate Zubairu, Ingrid Whiteman, Cath Bowden, Anna Nelson and Lisa Cherkassky. Other colleagues have also contributed to this book in their unfailing support for me and this project. Notably Anna Verges, Catherine Stanton, Caroline Bowsher, Fiona Ulph, Hannah Cobb, Ang Davies, Javier Garcia Oliva, David Lamb and Fiona Smyth.

A very special mention goes to Yiotta (Panagiota) Nakou. Yiotta is a PhD student with me but more importantly a treasured colleague and friend. Yiotta provided such amazing support to me during the writing of this book. At one point, I know she thought, as I did, that it might never come together. Her help, encouragement, generosity and wisdom were instrumental in getting this book over the line, and as a result, I will be forever grateful to her and look forward to many other shared projects in the future.

Special thanks also go to Tamara Montrose. Tamara worked with me on the FutureLearn course 'Introduction to Medical Ethics: The Impact of Disability Screening'. Her enthusiasm for this project encouraged me to develop the idea for this book. Tamara also helped me with the flow diagram that appears in this book, when trying to make all those arrows work on a page was threatening to drive me over the edge.

Unfortunately, my mum, dad, grandma and grandpa didn't live long enough for me to thank them here for all they did for me. My mum, Lorna Dirveiks, was the first of our extended family to go to university when she completed her MA in Education in the 1980s. I am sure her determination and the example she provided played a big part in any success that I have had in my own career.

Talking of family, I also want to thank my sister Beverley Bennett and stepfather Neil Dirveiks. It has been a difficult few years and your continuing support means a great deal to me and Finn.

To our Swedish friends – Eva-Lo, Ann, Anders, Stefan, Stella, Norma and Vega – thanks for the good times and the *slickepotts*!

Thanks also go to Gemma Grimshaw and Leanne Tuite for all the DL teamwork, to Jeannie for making my son a very happy man, to Sylvie for accepting us into your family and to Chorlton Sings Choir for the friendships, uncomfortable warm-ups and actions and the joy of belonging, with particular mention of the Disco Dollies. Thanks to friends, including Joanne and Gene Kerr, Sarah Flaherty, Alan Stowell, Alex Curry, George Oghani, Clare Pritty, Emma Fielden, Michael Mayall, Lara Ratnaraja, Leanne Mantack and Claire Elford-Hewitt.

Thanks to Alexander Bell my commissioning editor at Bloomsbury for your unwavering patience and positivity. Thanks also go to the anonymous reviewer of the draft of this book. All the mistakes that remain are mine, but the feedback you gave was so very helpful.

If you would have told my younger self (a) that one day I would be a professor, (b) that I would write a book like this and (c) that I would have so many wonderful people in my life to be thankful for, I would have been amazed. While writing a book and having an academic career are very important to me, it is the people who are in my life that I appreciate and treasure the most.

# PROLOGUE

I've been teaching and researching in bioethics and medical ethics for thirty years. I completed an undergraduate degree in philosophy in 1990 but then pursued my dream of teaching by training as a primary school teacher. However, in 1993, after only one year of teaching children in Year 6, an advert caught my eye for a role as a researcher at the University of Manchester with Professor John Harris, whose work I had admired as an undergraduate. Not knowing that primary schoolteachers with only a first degree do not usually have a chance of getting a research job at a prestigious university, I applied anyway. I was interviewed by Professor John Harris and two others of the founding members of our centre, Professor Margaret Brazier and Professor Anthony Dyson. During the course of the interview, I started to realize that they were expecting someone much more accomplished than me. The CV of the person interviewed before me, which was on the coffee table in front of me, confirmed that my competitor had not only a master's degree but also a PhD and a string of publications. I left the interview despondent and slightly embarrassed that I had put myself through this experience in front of these professors. When I got the call to say I had got the job, I couldn't quite believe it and seriously considered turning it down. I could never have imagined that thirty years later, I would still be working at the university and have established my own career in this area.

Understanding my own journey in bioethics might help you to understand why writing this book was important to me. After thirty years of grappling with issues around disability and reproduction, I wanted to write something that would help others to navigate this complex ethical area. Underlying regulation based on the welfare or best interests of future children are complex philosophical questions about how we can assess the quality of lives that have not yet started. This book draws on my decades of experience with the aim of enabling others to understand and engage with the issues that underlie regulation in this area and in doing so allow researchers, students, policymakers, service users and others to navigate this complex area of bioethical debate with confidence. My hope is that this book will give others the ammunition, tools, insight and impetus to evaluate existing regulation in this area and come to conclusions that will help to develop approaches to these areas of private life that are built on robust, reasoned argument rather than intuition and bias.

I have also written this book to empower others to gain confidence in tackling bioethical questions more generally. Questioning social norms, biases, established scholars and policymakers is often a daunting task. However, if we wish to build a society based on positions that can be defended and justified with confidence, then we must not only be committed to questioning existing regulation in these ethically controversial areas but also be committed to examining and questioning our own response to these ethical questions. This book suggests a way of doing this that will enable a thorough and critical approach to any bioethical question.

# INTRODUCTION

Could it be that our laws and policies around disability screening are based on bias and prejudice rather than reason and compassion? Might it be possible that regulation[1] that aims to prevent the use of fertility treatment in cases where there is concern for the welfare of these children is not only ineffective but also unjustifiably discriminatory, encouraging decisions based on subjective and often biased judgements?

In the not-very-recent past, regulation that prevented equal rights for particular groups in society such as women, people of colour and people from the LGBTQI+ community was a well-established part of the regulatory framework of many countries. While there was opposition to these regulations from individuals and groups in society, the general attitude towards these regulations, particularly from those with influence and power, was that these regulations were justified and necessary.

In this book, I argue that many of our current approaches to regulation around reproduction internationally, while well meaning and widely supported, are comparable to these historical regulations that turned out to be unjustified and motivated by bias rather than reason. I suggest that our regulatory approaches to screening for disability and to controlling access to fertility treatment based on concerns about the welfare of future children are not effective in protecting the welfare of any resulting children, undermine individual reproductive choice unjustifiably, cause harm and offence to many individuals and groups in society, and are based on bias and unexamined intuition.

You do not need to agree with me on this, but it is important, if we are to avoid the historical mistakes made around the rights of other groups in society, that we examine these issues in detail and find reasoned arguments for the regulations we endorse in the area of reproduction. If there is any possibility that these regulations around reproduction, disability and notions of what is a minimally acceptable quality of life may be another instance of unjustified and unjust regulation based on the cultural norms and biases of one group, then we have a duty to explore this possibility in detail. This is what this book aims to do.

## What do I mean when I talk about disability?

Before we start, it is important to briefly explore the language that I use around disability. What we mean by disability and even the appropriateness of using the word 'disability' is a complex issue. The language around disability is constantly evolving, and there will be varying language preferences among disabled people.[2] My aim in this book is to use non-stigmatizing and appropriate language around disability. To recognize the difficulties with a medical model of disability, I refer to disability within the social model of disability to acknowledge that 'the oppression and exclusion that people with disabilities experience are related to environmental factors, cultural attitudes, and social biases that influence how disabled people participate in society, and not merely a result of their impairments'.[3]

There is widespread disagreement around and preferences for whether the language used around disability should be 'person first' or 'identity first'. A person-first approach 'emphasizes distinguishing the person from the disability by referring to those with disabilities first as individuals and then mentioning their disability second and only when needed'.[4] An example of person-first language would be 'person with a disability'. However, there are those who take what is known as an 'identity-first' approach when it comes to the language around disability. An example of identity-first language would be 'disabled person'. It has been suggested that an identity-first approach has been adopted more recently by 'those who identify as disability rights advocates' as a 'manifestation of disability pride'.[5] There are, of course, many people who have what others might consider to be disabling conditions who do not identify as 'disabled people' or 'people with disabilities', and it is, of course, important to respect how individuals refer to their own identity. With all of this in mind I will take an identity-first approach to language around disability while recognizing that this will not accord with everyone's preference and identity.

## Where did my interest in all this start?

Very early on in my career, I was teaching a class with John Harris when he started to talk about his controversial claims that we have a moral[6] responsibility to eradicate disability. He asked the class to imagine that there were two hypothetical worlds, World A and World B, both containing the same number of human people. In World A, screening is so advanced that all disabilities have been eradicated. In World B, screening is less advanced and, while children are no longer born with severe disabilities that might make their lives a harmful experience overall, there are children born with conditions such as deafness, Down syndrome, autism and other conditions considered to be disabilities but not so severe conditions as to make their lives an overall negative experience. So in Harris' thought experiment,

- There are the same number of people in World A and World B.

- No one is born in either world with conditions that would be likely to make their lives unbearable.

- World A does not contain any disability, but World B does contain conditions such as deafness, Down syndrome and so on.

Despite both worlds having the same number of people who are likely to value their own lives, Harris argued that World A, where no disabled people are born, is an *ethically preferable* world and one we should be striving towards. This idea that we have a moral obligation to choose World A or choose to bring to birth the 'best' children possible is one that was later given the label the Principle of Procreative Beneficence[7] and fits with many people's intuitions about disability and reproduction. It is also one that, as we will see, has had a significant influence on regulation when it comes to possible reproductive choices. However, I found it really difficult to accept these arguments. I struggled with the idea that we might be able to agree on what is the 'best' child possible and, perhaps more fundamentally, that we might have moral reasons to eradicate even those conditions that are usually completely compatible with a life that is valuable to those who experience it.

For me, allowing individuals to make their own choices about reproduction seems a fundamental part of what allows humans to flourish. Attempting to influence or even restrict reproductive choices is therefore something that needs robust justification, but to my mind at least, this justification is not available. It concerned me that while I was not convinced about arguments around this responsibility to bring to birth the 'best' child possible, it was these same arguments that gave legitimacy to a great deal of existing regulation worldwide. My concern was that if we cannot provide strong reasons for this idea of a responsibility to bring to birth the 'best' child possible, then there is a danger that international regulation in this area is based on social norms, intuition and even bias rather than robust reason and argument.[8]

Over the past thirty years, I have explored these arguments that we have a moral obligation to bring to birth the 'best' child possible and my concerns about regulation based on the welfare of future children. I've written papers and given presentations that have explored the ethical foundations of these arguments and regulations in order to investigate and justify why it is that I feel uncomfortable about these regulations and the arguments that give them legitimacy. However, convincing others of my concerns here has not always been straightforward for two main reasons. Firstly, many of us share intuitions that preferring to bring to birth children with what we might consider the 'best' lives is the right thing to do. As a result, regulations that aim to avoid the creation of children with conditions or in conditions that are considered by many to be harmful fit with these intuitions and preferences and intuitively 'feel right'. Consequently, we are less likely to delve into the ethical foundations that justify these regulations. Secondly, the ethical foundations of these regulations require us to address complex philosophical questions, including questions about the value of human life, whether it is ever wrong to reproduce, what makes a human life valuable and whether we can compare the value of different human lives. The complexity of these questions and the concepts that they require us to explore mean that weaknesses in the ethical foundations of these regulations, if they exist, may not be easy to identify.

In writing this book, I wanted to use the work that I have done in this area to explore these issues further and to provide a guide to the various arguments and perspectives that underlie these regulations in detail. In doing so, I illuminate the concerns I have with these arguments and regulations and aim to enable you to examine and develop your own personal positions on these issues. You might not agree with my conclusions but whether you are a policymaker, healthcare professional, student, academic or prospective parent, this book will guide you through this debate and help you to develop clarity in your thinking around these complex issues, so that you can be confident in the conclusions you reach and your ability to defend them if necessary.

## Difficult decisions based on the welfare of the future child: Access to fertility treatment (including pre-implantation genetic testing)

Concern for the welfare of future children is understandable and something that healthcare professionals often feel very strongly about, particularly when they are instrumental in either enabling individuals to reproduce or providing services that aim to maximize the chances of individuals having the 'healthy' baby that is typically cited as the fundamental goal of human reproduction. As a result, when it comes to reproduction by in vitro fertilization (IVF) or other assisted reproductive techniques that are provided by healthcare professionals, the need to consider and safeguard the welfare of the resulting child seems an unquestioned requirement.

However, while many of us share the intuition that we should be concerned about the welfare of future children, particularly in areas like access to fertility treatment where healthcare professionals are instrumental in enabling reproduction, in practice, making decisions about projections of the welfare of future children is fraught with difficulties. Working in this area since the early 1990s, I have been aware of how difficult it is to make these decisions based on the welfare or best interests of children who do not yet exist, that is, future children. Numerous cases concerning difficult decisions made based on the welfare of future children have been brought to my attention by healthcare professionals, either seeking advice or raising ethical issues as part of training I have provided. As such, I have had somewhat unique access to cases that raise concerns about the welfare of future children but are invariably discussed behind closed doors to keep the all-important confidences of those involved.

While publicly reported cases of concerns around the welfare of the future child are relatively rare when it comes to access to fertility treatment, in my experience, these cases do present themselves fairly regularly and can take a significant amount of time and effort to resolve. Below are some examples of the sorts of cases I have been aware of over the years that have raised concern when it comes to individuals and couples attempting to access fertility treatment:

- Individuals who are wheelchair users (or have other conditions that may result in significant physical limitations) and there is concern about them being able to cope with the physical demands of parenting

- Individuals seeking treatment who have learning disabilities, autism or obsessive compulsive disorder where there is concern about their ability to cope with the challenges of parenting

- Cases where it becomes known that prospective parents have past convictions that may or may not involve violence, where suitability for parenting is questioned

- Couples where the male partner is in prison, again raising concern about the suitability for parenting
- Concerns around the choice of gamete donor
  - for example, there might be concern about a very close or blood relative being used for donation of sperm (e.g. the father of the woman wishing to become pregnant)
  - where the sperm donor is congenitally deaf and there is a concern about the resulting child being deaf
  - or a transgender woman who wishes to donate sperm before chemically transitioning but there is concern about whether there might be a genetic link to being transgender and thus concern about the welfare of any resulting children
- Concerns about the circumstances of gamete donation, for example:
  - where there is concern that the parents will not be honest with the child about the use of gamete donation, and this will impact negatively on the mental health of the resulting child
  - where there is concern that a lack of openness about the use of gamete donation might risk incest or other issues
- A couple who both have achondroplasia and wish to use pre-implantation genetic testing to select for an embryo with the same condition as themselves
  - Their motivation is that coping with an average height child would be challenging for them given their own height and that they live in a house specially adapted for the needs of smaller people. However, this raises concerns about selecting for a condition seen by many as disabling.
- A couple where there is suspicion of alcohol abuse – for example, clinic staff smelling alcohol on an individual's breath when attending clinic
- A couple where one individual has a history of mental health issues such as depression or personality disorders
- A couple where one partner is HIV positive
- A couple from the Deaf community whose first language is British Sign Language who wish to use pre-implantation genetic diagnosis to allow them to have a child who is also deaf, similarly raising

questions about using technology to select for a condition seen by many as disabling

- A couple where one partner has previously had a child officially removed from their care in order to protect that child from potential harm

This is clearly not an exhaustive list but represents the sort of cases and concerns that have been and continue to be discussed as part of access to fertility services. All the above cases are based on real cases that I have encountered through my contact with those working in clinics. While cases that raise concern and result in investigation are thought to be relatively rare, these cases do exist and cause a great deal of concern and scrutiny of individuals as part of their already stressful attempt to become parents.

The cases above all arose from individuals and couples attempting to access fertility treatment and pre-implantation genetic testing, and this is where we see many of the very clear instances of attempts to assess the welfare or best interests of future children. However, this notion of concern for the welfare of future children is the basis for current regulation in many areas of reproduction, both in the UK and internationally. I argue that it is also this concern that justifies a different approach to informed consent that is evident in routine prenatal screening programmes for conditions like Down syndrome. I argue that routinization of these screening programmes necessarily introduces an element of coercion. This different approach to informed consent may be justified as an attempt to prevent harm to future children by encouraging a high uptake of screening. As a result, this notion of concern for the welfare of future children impacts a great many areas of the current regulatory framework in the area of reproduction worldwide and thus has the potential to affect the reproductive choices of a great many individuals.

## The future of screening in reproduction?

Considerations of the welfare of future children have become an established part of regulation in the area of assisted reproduction, pre-implantation genetic testing and prenatal screening internationally. The use of assisted reproduction and the scope of pre-implantation genetic testing and prenatal screening are likely to expand greatly over the next decades.

With people delaying pregnancy and the trend for egg freezing increasing, reproduction by IVF continues to grow, and it is even suggested that in the near future 'as many as 10% of all children will be conceived through IVF in many parts of the world'.[9] As the use of IVF as a means of reproduction increases, so will the opportunities to scrutinize the conditions that these new lives will be born into. A requirement to consider the welfare of the future child when it comes to access to fertility treatment will have the potential to impact on many more reproductive choices of individuals seeking this treatment.

Our attempts to maximize the number of 'healthy' children born each year have resulted in the development of pre-implantation genetic testing that currently allows IVF embryos to be screened for genetic conditions where there is a history of these conditions in a family or other reasons to be concerned about the presence of these conditions. Routine prenatal screening of pregnant people[10] was introduced for similar reasons, to enable individuals to choose to avoid bringing to birth a child with certain conditions such as Down syndrome.

The development of whole-genome sequencing (often referred to by the abbreviation WGS) means that we can now gain a huge amount of information relatively quickly and cheaply about any human being's genetic make-up, whether this is an adult person or a pre-implantation embryo or foetus during pregnancy. As a result, there are calls to substantially expand the scope of pre-implantation genetic testing by using WGS.[11]

The development and introduction of non-invasive prenatal testing (often referred to by the abbreviation NIPT) to routine prenatal care presents an opportunity to expand screening using WGS in pregnancy. NIPT involves screening and testing genetic material found in the pregnant person's blood sample and thus involves no risk to the foetus. Using this foetal genetic material, it will be possible to sequence the whole genome of the foetus during pregnancy and to identify all sorts of genetic conditions and predispositions at this stage in development. This ability to expand prenatal screening dramatically has led to calls to introduce this new approach more widely.[12]

We can quite easily imagine a time in the not distant future where WGS is used to screen all IVF embryos for any kind of genetic condition that might be seen as negative to allow the choosing of the 'best' embryo to be brought to birth. Anything that is seen as an impairment could be avoided in this way, and this could extend to so-called non-medical traits such as height and intelligence and even conditions such as autism and other similar conditions if a genetic marker is identified.

The possibility of using WGS in routine screening in pregnancy seems to simply extend and enhance the current policies of trying to reduce the number of children being born with conditions that are seen as disabling. Presenting these extensive screening programmes in a way that aims to maximize uptake of screening seems to fit in with our already established approach to prenatal screening.

While expanding the screening of pre-implantation embryos and in pregnancy has the potential to provide a great deal of information, not everyone will wish to have this information. This may be because information provided may relate to conditions that are not universally seen as negative or because the information provided is inconclusive or relates to adult-onset conditions or even carrier status.[13] This huge expansion of screening is on the very near horizon and will require us to develop a regulatory framework that is both ethically justifiable and practically workable. The obvious approach to the expansion of screening in this area is to simply extend the regulations we currently have that focus on the welfare of resulting children.

However, in this book, I argue that if we were to base future regulation in this area on our current regulatory framework that focuses on the welfare of resulting children, future regulatory frameworks, like our current ones, will be complex and well meaning but, ultimately, practically unhelpful and without the strong ethical foundations that we would wish regulation to have. Thus, while this book focuses on the *current* regulatory framework that governs reproductive choices based on the notion of the welfare of future children, I argue that the considered conclusions we reach about these *current* approaches should and can provide guidance when it comes to the regulation of these *future* developments in the screening of future children.


## Introducing three questions around screening and reproductive choices

In order to explore and evaluate the ethical foundations behind regulation in this area, in this book, I focus on three questions that represent three distinct areas where regulation is often based on concern for the welfare or best interests of future children. These three questions are:

- Question 1: Are we justified in attempting to evaluate the potential parenting ability of those trying to access fertility treatment (e.g.

disabled people or individuals with past criminal convictions) and prevent access in some cases?

- Question 2: Should we allow prospective parents using IVF to implant[14] an embryo with a condition considered a disability? For example, should a deaf person be allowed to implant a 'deaf' embryo?
- Question 3: Is routine screening for Down syndrome in pregnancy ethically acceptable even if there is evidence that individuals may feel pressure to accept this screening?

> What are your initial thoughts on these questions? You might want to make a note of your thoughts somewhere so you can return to these.

Most of you will have a strong reaction to these questions and the scenarios they come from. When the sorts of cases represented by these questions hit the headlines, it is clear that many of us find answering at least some of these questions deeply troubling.

When faced with these questions, many of us are concerned about the choices that these prospective parents want to make and are concerned about the welfare of any child who will result. This sort of response fits well with a general instinct that most of us have to protect children wherever possible and give them as many positive opportunities as we can. Many of us would choose that our children and other children brought to birth are not disabled, and the birth of a disabled child is often seen as something that parents might regret and that as a society we may wish to avoid. Assisted reproduction providers are often concerned about whether they should be providing access to fertility treatment to prospective parents who face possible parenting challenges such as medical conditions, disabilities or social or mental health issues. In all these cases, this unease is focused on a concern for the welfare of the child who might result from the chosen actions of prospective parents to bring a child to birth in what many might see as suboptimal circumstances. Unease about these cases is based on concern for the welfare of the children who will come to birth as a result of what are often seen as unwise or ill-advised choices.

While it may seem uncontroversial that we should be concerned for the welfare of any children who may result from these choices, finding adequate justification for regulation based on this notion of the welfare or best interests of the child is difficult to pin down. For several decades the use of this notion of the welfare of the future child to regulate reproductive choice has been questioned.[15] There is a central reason why, for many, the legitimacy of this focus on the welfare of the child is undermined when it comes to the creation of future children. To begin to understand this think about your answers to the following question:

> Do you think being alive is generally a good thing for most people?

We live in a society that generally takes the answer to this question for granted. We often hear people talk about feeling lucky to be alive, we celebrate the birth of a new human life and grieve the ending of other lives. As a result, we often take it for granted that being alive is generally a good thing. But as we will see later in this book, while this is the predominant view in society, there are those who do not share this general view that being alive is generally a good thing and argue that having children benefits no one and causes those born to suffer unnecessarily.[16] I suggest that deciding where we stand on this question, of whether being alive is generally a good or bad thing, is fundamental to our approach to the ethical questions we explore in this book and the first fundamental question that you need to come to a position on.

If you do *not* think that being alive is generally a good thing, then this position would seem to lead us to a conclusion that all reproduction should be avoided if possible. Further if you take the view that all lives are unacceptably bad then it would be difficult for those holding this position, to justify any assisted reproduction. This pessimistic view of human life would also make it hard to justify prenatal screening for particular conditions if the welfare of all new life is considered to be unacceptably low.

However, most of us do not take this position and hold a position that being alive and creating new life is generally a good thing. For those of us who hold this position we are left with further questions to consider when thinking about the welfare of future children:

- *If you think that being alive is generally a good thing for most people, do you think that there are lives that are not good for the people who experience them, that is, where the good of life is overwhelmed by suffering?*

- *How bad does something have to be to overwhelm the good things in life?*

- *Do you think being born deaf is something that would be likely to overwhelm the good things in life?*

- *Is being born with a condition like Down syndrome something that would overwhelm the good things in life?*

- *Would being born to parents with disabling conditions or past convictions be likely to overwhelm the good things in life?*

These are complex and difficult questions but ones that we must address if we are to arrive at justifiable regulation in this area. While our intuitive reactions to providing unchecked access to fertility treatment or to someone wishing to use pre-implantation genetic testing to select for disability as part of IVF might be a significant concern, as we will see, if we think that being alive is a good thing generally, then it is difficult to find compelling reasons why we should interfere with the reproductive choices of individuals on the basis of the welfare of a future child. While those of us who may not have had direct experience of those living with conditions such as deafness or Down syndrome may have a negative perspective of the quality of life of those living with these conditions, if we listen to disabled people themselves then the picture is very different. In fact, disabled people are as likely as anyone else to value their own lives. As a result, if we take the view that life is generally a good thing, we might conclude that when a child is born in the only condition/s they can be born in and with a life they are as likely as anyone else to value, it is difficult to understand who has been harmed by this choice. However, this conclusion, that no one has been harmed by this choice and thus that we may not have good reasons to interfere with these choices, is, for many very counter-intuitive and goes against their natural instincts to attempt to protect the welfare of future individuals.

Although I have tried to introduce these arguments in a simple way here, these are extremely complex arguments but ones that underlie a great deal of regulation in this area. I argue that it is imperative that we understand these underlying arguments and the questions they raise in some detail to ensure that we can develop regulation regarding reproduction that can be ethically justified. This book aims to outline and explore these arguments and questions in a way that enables this in-depth understanding of these often-complex issues.

Regulation around reproduction has the potential to have a significant impact on the lives of pregnant people, prospective parents and those living with the sort of disabling conditions that are often the focus of screening. Understanding these complex ethical issues, while challenging, is paramount for anyone involved in this area of reproductive choices whether that is individuals and groups who are tasked with developing a regulatory framework in this area, healthcare professionals who work within this framework, ethics committees who support decision-making in this area or prospective service users[17] wishing to access clinics services. To arrive at a position on these issues that you can be confident of you will need to navigate complex philosophical questions such as:

- *Do we ever harm individuals by bringing them to birth?*
- *Is it ever morally wrong to reproduce?*
- *Do we have a moral obligation to try and bring to birth the 'best' children we can?*
- *How do we make decisions in this area when there is no consensus regarding what is the morally right thing to do?*

In this book, I explore the current approach to reproductive choices based on consideration of the welfare of the future child, the arguments that underlie these approaches and the questions that these approaches raise. There is, of course, no consensus about what the 'right' answers are to these questions. However, despite this, answers are needed if we are to develop current and future regulation in this area that we can defend. In this book I show how these arguments and questions can be broken down in a way that will help anyone facing these questions to get to the heart of the issues involved. I support you to come to your own positions on these questions, providing an ethical toolkit that that encourages the development of well-reasoned, unbiased answers to these often-complex ethical questions. In doing so you will be able to examine the ethical foundations of our current regulatory approaches to reproductive choices around future children and come to a view about whether our current approaches are the right ones and the right ones for future regulation in this area.

# CHAPTER 1
# INTERNATIONAL REGULATION AROUND THE WELFARE OR BEST INTERESTS OF FUTURE CHILDREN

When it comes to the ethics of reproduction, one thing that seems uncontroversial is that when a choice is made that will affect the life of a child, then considering the welfare or best interests of that child should be fundamental to this decision-making process. This consideration of the welfare of the child is reflected in regulation globally and in international conventions such as the 1989 United Nations Convention on the Rights of the Child, Article 3(1), which states 'in all actions concerning children, whether undertaken by public or private social welfare institutions, courts of law, administrative authorities or legislative bodies, the best interests of the child shall be a primary consideration'.[1] This concern for the welfare of any children affected by actions and choices guides a great deal of regulation when it comes to existing children – for instance, around issues such as adoption, healthcare and child protection.

It is perhaps unsurprising then that this principle is often also highly influential when it comes to regulation around reproductive choices that will affect who will be born. Unlike regulation and policies around adoption or child protection of existing children, these are based on concern for future children – that is, those who are not yet born.

## What do I mean by future children?

In this book, we are focusing on regulation that is based on concern for the welfare of children who have either not yet been conceived (in the case of regulating access to fertility treatment), or embryos that have not yet been implanted (in the case of pre-implantation genetic testing), or foetuses in the womb that are subject to prenatal screening but have not yet been born. For many people, including myself, the ethical issues around these decisions are very different from the ethical issues around decisions we might make about children who have already been born. Those who take this view argue that in these cases, the children who will be created as a result of these decisions do not yet exist in the same way that children who have already been born exist. This is why, in this book, I refer to these sorts of decisions as decisions about the welfare or best interests of *future* children.

However, it is important to recognize that this view about all of these decisions being related to future children, that is children who do not yet exist, is one that is based on a particular stance on the moral status of the embryo and foetus and this particular stance is not one that is shared by everyone. Most views on the moral status of the embryo and foetus would agree that when decisions are taken before conception, these are choices about future children and not about an entity that currently exists. However, when it comes to pre-implantation embryos and foetuses that are screened in pregnancy, this general consensus ends.

The debate around when human beings start to exist as entities that have interests in self-preservation – and thus that we have an obligation to provide protection to – is one that has run for thousands of years. There are many and varied positions on this issue. There are those, for instance, who argue that the embryo should be treated as having the same status as you or I, from the moment of conception. Others argue that it is sentience that marks the moment in development that this entity has begun to exist in the sense of becoming a being with an interest in survival. Still others will argue that while the embryo or foetus is inside a pregnant person's body that person's interests prevail, but once a foetus has developed to the stage that they can survive outside a pregnant person's body, then we should make decisions about the foetus as if they have the same status as more fully developed children and adults. These are only a few of the many stances on this question of the moral status of the embryo and foetus. Coming to your own position on this issue is really important when it comes to considering questions around the ethics of reproduction, as the particular view you take on this issue will have a big effect on where you stand on reproductive issues.[2]

I do not have space in this book to explore this complex and wide-ranging area of bioethics comprehensively enough to do justice to this debate, but I can explain my own stance on this issue and why it is that, in my view, embryos and early foetuses are considered to be *future* children and not considered to have the same status as *actual* children.

The stance I take on this tricky issue is that it is the development of characteristics like self-awareness and the ability to value one's own existence that marks the beginning of a particular human biographical life. These cognitive characteristics develop either in very late pregnancy or early infancy. As a result, I and others taking this view of moral status (often known as the personhood view) would argue that the embryo and early foetus have the status of *future* rather than *actual* children at this stage in their development.

This view that the embryo or the foetus (at least until the third trimester of pregnancy) does not yet have intrinsic interests means that this stance is compatible with allowing abortion and other practices such as IVF (which inevitably involves the destruction of 'spare' embryos that are not implanted). While my view is that embryos and foetuses are not yet existing persons, where there is an intention to bring these entities to birth and thus enable them to develop into a self-conscious entity with interests in their own welfare, it will be important to ensure that decisions made in this early stage do not impact negatively on the children these embryos and foetuses may become. To clarify this further, on my view, the embryo or foetus is not harmed if it is never brought to birth as, at this stage, this entity is not able to value its own existence and thus is not denied something it values. However, while it may be permissible not to bring an embryo or foetus to birth, it is still possible to harm a person who will come to exist by decisions made before they are born, either by damaging their welfare at this early stage (perhaps by taking drugs that cause damage to the foetus) or allowing them to be born in conditions that fall below a threshold that is acceptable in terms of their own quality of life.

I respect the views of those who take a different view of the moral status of embryos and foetuses, that is that they should be treated as having similar status as existing children before the third trimester of pregnancy. However, those who argue that embryos and/or early foetuses should be treated in the same way as those who have been born are unlikely to argue in favour of allowing access to abortion or fertility services such as IVF, which inevitably entail the destruction of some embryos. As a result, questions about how we decide who should access fertility treatment, including the use of pre-implantation genetic testing, and whether we should endorse prenatal screening that involves an element of coercion towards termination of pregnancy seem to be clearly answered by these stances on the moral status of embryos and early foetuses. If we accept that the embryo or the foetus in the first two trimesters should be treated in the same way as you or I, then any practice that involves the inevitable destruction of these entities would seem unacceptable from this viewpoint. As a result, this stance seems fundamentally incompatible with our current regulations on abortion and the availability of IVF.

While access to abortion and IVF are justified for those of us that take the view that the questions we are engaged with in this book focus on lives that have not yet started, this does leave us with a responsibility to engage with the difficult questions that I have raised in this book about how we make sense of decisions around the welfare of future children, that is children that at the time of the decision-making do not yet exist. When I talk about decisions that will affect future children, then these are the sorts of cases I am considering.

## Regulations based on the welfare of future children

As you have seen, the three main questions we will focus on in the book raise examples of regulations that concern the welfare of future children. These regulations are as follows:

- Controlled access to assisted reproduction treatment (IVF, donor insemination, etc.)
- Prohibition of the use of IVF embryos known to have conditions such as deafness
- Routine screening programmes for Down syndrome in pregnancy which aim to encourage a high uptake of screening.

I will argue that these seemingly disparate regulations are united by their underlying aim to influence reproductive choices, with the goal of protecting the welfare of any resultant future child. As we will see, the result of these regulations is that some reproductive choices are discouraged or even denied on the basis of this concern for the welfare of the child who will result from these choices. Underlying this regulation is the idea that we have a strong moral obligation to attempt to prevent the birth of children in what are seen as suboptimal circumstances, and this idea has a great deal of public[3] and academic support.[4]

In this book, we will explore this regulatory framework that focuses on reproductive choices and, in particular, regulation that extends concern for the welfare or best interests of existing children to children who do not yet exist. I focus on the regulatory framework in the United Kingdom (UK), but the analysis I provide has international relevance as a great many other jurisdictions have similar regulation, often based on the UK approach.[5] Before we move on to the exploration of the ethical arguments and questions that underlie these regulations, I first want to outline what these regulations are and how they focus on the idea of the welfare of future children.

## Screening prospective parents attempting to access fertility services

The area of regulation that immediately comes to mind when we think about the welfare of children is the screening of prospective parents who are attempting to access adoption, fostering or fertility services. Screening individuals who reproduce naturally is impractical and prohibitively invasive. However, where outside agencies are involved with creating parents through adoption or fostering, it seems obvious that this role should include a responsibility to ensure that it is the best interests of the child that drives these decisions. As a result, there are important, and often extensive checks, made on those who wish to adopt or foster existing children with the aim of choosing a home environment for this child that maximizes their welfare.

For many it will seem important, for similar reasons, to extend these checks and evaluations to those seeking help to start a family using assisted reproduction, including IVF. As Guido Pennings points out,

> Since we can control (at least to a certain extent) the circumstances in which a child is made when the candidates are infertile, we ought to restrict our co-operation to those cases which maximize the welfare of the child. At the same time this fact explains why the standard for medically-assisted procreation must and can be higher than for natural reproduction.[6]

This strong feeling that we should attempt to safeguard the welfare of any children born as a result of fertility treatment has meant that many countries that offer these services not only consider any physical or medical challenges that might face a child born as a result of these services, but also include an assessment of social factors which may indicate the prospective parents' ability to provide a suitable home environment for any resulting child.

What is often viewed by other countries as a '[a] model for best practice'[7] in this area is found in the UK's Human Fertilisation and Embryology Authority's (HFEA's) Code of Practice. This Code of Practice takes its lead from the Human Fertilisation and Embryology Act 1990 (as amended in 2008) to provide further guidance in assessing and, in some cases, refusing treatment for licenced fertility treatments based on the suitability of prospective parents for treatment in terms of the best interests of this future child. This so-called Welfare of the Child provision states that

> 8.2 The centre should have documented procedures to ensure that proper account is taken of the welfare of any child who may be born as a result of treatment services, and any other child who may be affected by the birth.
>
> 8.3 The centre should assess each patient and their partner (if they have one) before providing any treatment and should use this assessment to decide whether there is a risk of significant harm or neglect to any child referred to in 8.2.[8]

As a result, all UK licensed fertility clinics are obliged by law to consider the welfare of any child born as a result of the treatment they provide. This results in individuals and couples, where there is a disclosed or observed reason for concern, being scrutinized by the clinic in an attempt to determine whether there is a 'risk of serious harm to any child'[9] as a result of this treatment. The wording of the Welfare of the Child provision allows a very flexible application of this requirement as the obligation is merely to take into account the welfare of any resulting children; it doesn't specify in detail what should be considered as harmful, what counts as 'serious harm' and what further actions should be taken if concerns are identified.

In the past, this provision was often interpreted to prevent single women or same-sex couples from accessing fertility treatment. But as attitudes have changed in the UK, and evidence has developed to demonstrate that the fact that someone will be a single parent or parent as a same-sex couple does not pose a risk to the welfare of their children,[10] this is no longer the case. However, this provision is still used to scrutinize the projected parenting abilities of individuals and couples who attempt to access fertility treatment. While these investigations and decisions are usually confidential, and thus gaining detail of the sorts of cases that are scrutinized in this way is difficult, a study reported in 2015 confirmed anecdotal evidence that the sort of thing that 'served as prompts for staff to make efforts (sometimes considerable ones) to find the evidence that would alleviate concerns' were '[d]rug or alcohol abuse, mental health issues, a disability, or a previous conviction involving a child'.[11] Other factors that might raise concern were the perceived lack of stability of the relationship, perception of financial hardship, no longer having custody of an existing child and suspicions of domestic violence.[12] There is evidence that certain categories of individuals may be investigated and that these investigations may involve gathering evidence from different sources, including the individual's GP, the probationary service, police and social services.[13] It seems very likely that such investigations may take some time and, while cases where patients are refused treatment entirely are difficult to find, that these protracted investigations may well result in patients becoming too old for treatment, giving up[14] or attempting to get treatment at another clinic where the Welfare of the Child provision may not be applied in the same way.[15]

The UK Welfare of the Child provision, as it relates to access to fertility treatment, has generally been seen as a model for regulation in other jurisdictions, resulting in a great many other countries taking a very similar approach to the regulation of fertility treatment.[16] Internationally, the existence of regulations addressing the welfare of the child are common, and in a survey from 2021, 63 per cent of seventy-three respondents (clinical staff members from assisted reproduction centres) reported that 'fertility treatments might be denied if significant concerns about the potential future welfare of the child were uncovered by the clinic staff',[17] although only 45 per cent of these respondents reported the existence of regulation that addressed the welfare of the child in this context. The existence of this regulation is more common in Europe, but examples of this approach are also reported in Africa, the Americas and Asia and Australia/New Zealand.[18]

These regulations are enacted through a mixture of federal or national law, state or regional law and professional guidance or standards. For example, the Australian Government's National Health and Medical Research Council published professional guidelines that state that 'ART [Assisted Reproductive Technology] activities should not commence without serious consideration of the interests and wellbeing of the person who may be born as a result of that activity' and that 'clinics may refuse or delay treatment (pending further review by the clinical team) if there are concerns about the physical, psychological and/or social wellbeing of any relevant party'.[19] In Canada, the Assisted Human Reproduction Act declares that 'the health and well-being of children born through the application of assisted human reproductive technologies must be given priority in all decisions respecting their use'.[20] Similarly it is reported that French law 'stipulates that it is the prerogative of the health care team to refuse access to ART[Assisted Reproductive Technology], on a case-by-case basis, when they consider for one reason or another that "the welfare of the future child" might be endangered or compromised'.[21] The Nordic countries of Denmark, Iceland, Finland, Norway and Sweden all have legal statutes that restrict access to fertility treatment explicitly on the basis of the welfare of the resultant child.[22] These widespread international restrictions, like their UK equivalent, invite those providing treatment to make an assessment about the expected welfare of children born as a result of fertility treatment and subject some individuals to a great deal more scrutiny than others and will also result in small groups of individuals not being given equal access to fertility treatment.

Around half of the European countries, including the UK and more recently France, no longer require those seeking fertility services to be in a stable heterosexual relationship to be eligible for fertility treatment. However, being in a heterosexual relationship is still requirement for access to treatment in a significant number of European countries, including the Czech Republic, Italy, Lithuania, Poland, Slovakia, Slovenia and Switzerland.[23] Internationally, it was reported that 54 per cent of eighty countries surveyed did require that individuals provided with fertility services 'must be in a stable heterosexual relationship and that this was documented by laws or statutes, oversight by professional organizations, or government agencies with relevant jurisdiction'.[24] The situation for transgender and non-binary individuals is also problematic in some areas, with a number of countries explicitly prohibiting access to fertility treatment for transgender individuals. Even where there is no explicit bar to access, there may be other barriers. For instance, some countries require sterilization for transgender individuals seeking recognition of their gender identity, whereas other countries do not recognize the gender identities of transgender individuals[25] or even criminalize diverse gender expression.[26] These restrictions mean that many individuals and couples who do not fit this idea of a heterosexual couple may struggle to access fertility treatment in their own countries and may lead them to travel in order to access fertility treatment.[27] It seems likely that what motivates these restrictions for same-sex couples, single women, and transgender and nonbinary individuals is a 'traditional' view of what the ideal environment might be for child rearing even if this motivation is not explicit. While these restrictions do not explicitly come under considerations of the welfare of the future child, it seems likely that these concerns about the welfare of resulting children motivate at least some of these restrictions.

While access to fertility treatment is the classic example of the use of regulation based on the concern for the welfare of future children, I argue that there is evidence of the influence of this principle in other areas of regulation around reproductive choice.

### Preventing selection for disability using preimplantation genetic testing

Perhaps the most strident example of these legal restrictions focusing on the welfare of future children can be seen in the UK approach to pre-implantation genetic testing (often referred to as PGT). Pre-implantation genetic testing, or to give it its full name pre-implantation genetic testing for monogenic or single gene disorders (sometimes referred to as PGT-M) was previously known as pre-implantation genetic diagnosis or PGD and involves testing an early embryo created outside the human body using IVF for genetic or chromosomal conditions. At the time of writing, in the UK, there are over six hundred conditions that this procedure can be used to detect,[28] and it is possible to add conditions to this list by application to the Human Fertilisation and Embryology Authority.[29]

In the UK, it is explicitly prohibited by law to implant embryos shown by pre-implantation genetic testing to have a condition that is viewed as a genetic 'abnormality'. This applies to 'a serious physical or mental disability', 'serious illness' or 'any other serious medical condition' that is considered to pose a 'significant risk' to the person who will result from implantation of this embryo.[30]

As a result, UK law is clear that any embryos identified using pre-implantation genetic testing as being at risk of developing such conditions 'must not be preferred to those that are not known to have such an abnormality'.[31] While I focus on the regulation with regard to pre-implantation genetic testing on IVF embryos prior to implantation, it is important to note that this UK legal prohibition applies to choice of gamete donors as well as IVF embryos. As a result, it is also illegal to prefer the use of gamete donors if this choice is likely to result in a significant risk of creating a child with conditions that might be considered to be serious physical or mental disabilities, illnesses or other serious medical conditions. The addition of this clause was included in an attempt to prevent individuals from deliberately selecting embryos with genes known to result in disability and is thought to have been introduced in response to a controversial case where a deaf lesbian couple chose a sperm donor who was also deaf in an attempt to choose that their child would also be deaf.[32] The choice of language in this legal clause on both implantation of embryos and choice of gamete donor makes it clear that the motivation of these restrictions is to protect the welfare of those who will be born from something that is considered unquestionably negative, with talk of 'risk' rather than a more neutral word such as 'probability' to the person who might result and of 'abnormality' rather than 'conditions'.

There is no indication of what is meant by a 'serious physical or mental disability' in the UK regulation, but given that the inclusion of this clause was motivated by an attempt to prevent individuals selecting for conditions like deafness, this is likely to relate to most other conditions usually seen as disabling. While it is unlikely that many prospective parents would undertake pre-implantation genetic testing only in order to select for a particular disability,[33] it can be imagined that there will be those undergoing IVF and pre-implantation genetic testing who do not wish to reject otherwise 'healthy' embryos with conditions such as deafness, Down syndrome, particular conditions that cause achondroplasia or other comparable conditions. This may be for all sorts of reasons, but one of these reasons might be that the prospective parents also live with these conditions and may not want to exclude the implanting of embryos who will share the same condition they live with. In most cases, the whole provision of IVF is focused on attempting to provide individuals with a chance to parent a child who is 'like them' as much as possible, with genetic relatedness highly prized and, where genetic relatedness is not possible, choosing gametes that mimic as far a possible a strong genetic connection between parent and child.[34] However, this UK legislation explicitly prohibits, for example, deaf individuals or individuals with achondroplasia from choosing to implant an embryo with the condition they have.

While there are some jurisdictions, including the United States and Mexico, that do not regulate the use of pre-implantation genetic testing[35] many countries take a very similar approach to the UK, allowing access to pre-implantation genetic testing for those at risk of passing on particular conditions but restricting its use to selecting *against* genetic conditions and diseases that are generally seen as negative. This approach is even seen in countries that previously prohibited pre-implantation genetic testing entirely, including Germany, Switzerland and Austria.[36] As a result, while there is often no clear Statute law in other jurisdictions prohibiting the implantation of embryos seen as impaired as there is in the UK, typically, the use of pre-implantation genetic testing is clearly targeted at selecting against disability and 'disorders'. As a result, in these regulated jurisdictions, it seems likely that prospective parents who are deaf or have another condition seen as disabling would be discouraged, if not prohibited, from choosing to implant a child with the same condition they have, particularly if there are other unaffected embryos available.

## *Routine screening for Down Syndrome in pregnancy*

Legal restrictions on access to fertility treatment and restrictions on the use of pre-implantation genetic testing based on the welfare of the resultant child are clear examples of regulation that is based on this idea of prevention of harm or the best interests of future children. However, there are, arguably, other, less explicit examples of practice and regulation that may be motivated in a similar way. In what follows, I suggest that routine prenatal screening for Down syndrome is one of these examples of a policy that is strongly motivated by what might be considered concerns for the welfare of future children.

As we will see, while I argue that a strong motivation for routine prenatal screening for Down syndrome is this concern for the welfare of the future child, it is a rationale around the enhancement of respect for reproductive autonomy by enabling people to make more informed choices about their pregnancy that is usually used to justify this approach to screening and testing for this condition.[37] While reproductive autonomy is undoubtably part of the motivation for these routine screening programmes, as we will see, their effectiveness in enhancing pregnant people's autonomy can be limited. The goals behind this routine screening are complex and also involve aims to prevent 'burdens' on society from the birth of those with Down syndrome. While these kinds of societal concerns may have been instrumental in the introduction of this kind of screening, justifying this motivation is highly problematic. It is for these reasons that I suggest that there is a strong case to argue that routine screening for Down syndrome is another example of a regulation that is strongly motivated by concern for the welfare of future children.

Down syndrome is a condition that's caused by an extra copy of chromosome 21. This condition is associated with particular facial features, and children with Down syndrome usually learn and process more slowly than children without this condition. Down syndrome is also often associated with higher rates of health issues such as congenital heart disease, obesity, sleep apnea and dementia. Individuals with Down syndrome typically have a lower life expectancy than those without the condition, but this is improving, and there is evidence that this may be improved further by improving the often-poor rates of preventive healthcare in this population.[38]

Some pregnant people will want to know whether the foetus they are carrying has Down syndrome, as they may decide that they want to end the pregnancy with a termination or they may find this information important when preparing for the birth of a baby with this condition. In the UK and many other countries, screening tests for Down syndrome have been offered to pregnant people since the 1980s.[39]

Typically, as part of routine screening in pregnancy, a screening test for Down syndrome is offered to all pregnant people between ten and fourteen weeks into their pregnancy. Screening for Down syndrome is offered in the same screening process as screening for Edward syndrome and Patau syndrome, but you can choose whether you wish to receive results for all three conditions, two conditions, one condition or none of these conditions. Since the 2000s, the screening offered has been a combined test that uses maternal age, the nuchal translucency measurement taken during an ultrasound scan and a blood test to calculate the chance of the pregnancy being affected by Down syndrome. This combined test is a screening test, as it doesn't give a definite diagnosis, but it will give pregnant people an indication of the likelihood that their foetus has Down syndrome. Further tests are needed to give a more accurate diagnosis. These further tests involve taking either a small sample of placenta or some fluid from around the foetus and testing it. These further diagnostic tests are more accurate and known as invasive tests, as a needle is used to collect the material to be tested; they also involve a small risk of miscarriage.[40] More recently, a new diagnostic test has been added to routine screening programmes around the world. This new diagnostic test is a non-invasive prenatal test (also known as NIPT) and involves testing for fragments of the foetus's DNA that can be found in the pregnant person's blood and, therefore, doesn't involve a risk of miscarriage.[41]

Routine screening for Down syndrome, Edward syndrome and Patau syndrome is accepted by the vast majority of pregnant people. For instance, in the UK in 2020–1, it was reported that 85.2 per cent accepted this routine screening involving the combined test.[42] The percentage of those who terminate after a prenatal diagnosis of Down syndrome in the UK has remained fairly constant between 2013 and 2017 at around 90 per cent.[43]

While routine prenatal screening programmes for Down syndrome are, in most cases, non-compulsory and do not prohibit reproductive choice entirely, there is a great deal of evidence that routinely offered screening programmes of these kinds do not adhere to the usual standards of voluntary informed consent[44] (as set out in the UK by case law). The legal requirements for voluntary informed consent in the UK and in other countries with similar regulations aim to ensure that medical procedures, including diagnostic tests, are only performed on competent adults who are sufficiently informed of what this test involves, including the risks and benefits, and who are not being pressured into accepting this test or procedure. This is done because we usually assume that, ethically and legally, people should have control over what is done to their bodies and thus to guard against medical paternalism.[45]

In order for consent to be legally and ethically valid, individuals need to have sufficient information to make a meaningful decision about screening, this decision should be voluntary, that is, without any obvious coercion, and individuals should be able to understand the decision they make. Only if these conditions are satisfied can pregnant people be said to have given valid legal consent and we can have any confidence that their choice to accept screening is their authentic choice.

While guidance on prenatal care emphasizes that pregnant people should understand that it is their choice whether these tests are done[46] and implies that the usual standards of voluntary informed consent should still apply in this setting, there is evidence that, in practice, these standards are often not met for routine prenatal screening of this kind. Studies consistently show that levels of knowledge adequate for consent to be considered sufficiently informed are not being achieved in a significant proportion of participating pregnant people,[47] because people are often not aware that screening is optional,[48] many do not know why they are being screened[49] and there is evidence that routinized screening of this kind puts pressure on pregnant people not only to accept screening and testing but to opt for a termination of pregnancy with a positive result.[50]

The main problem here is that the routine nature of these screening programmes may not be compatible, in many cases, with gaining voluntary informed consent. Routine screening programmes are implemented in order to gain a higher uptake of screening than simply offering this screen and test in the way that we would normally offer other genetic tests as an elective 'opt in' test. Prenatal screening for Down syndrome has become a routine, established and accepted part of prenatal care around the world. It has been suggested that the reason there is broad acceptance of routine prenatal screening of this kind is that it was added as 'just [another] simple blood test' to existing prenatal care, which was a well-established part of pregnancy, that was accepted as a way to prevent harm to the pregnant person and the future child.[51]

Making Down syndrome screening a routine part of prenatal care with the aim of getting a high uptake of screening sends the message that this screening is important and recommended. There are concerns that this idea of recommendation puts pressure on pregnant people to participate in screening and that as a result '[i]t could be difficult for women to decline without appearing irresponsible and blameworthy.'[52] The fact that screening is often seen as 'just a blood test' and one that is offered by 'trusted' medical professionals means that pregnant people can feel that they should accept screening and that by doing so they are being responsible parents.[53]

The introduction of the more recent non-invasive prenatal test into routine screening programmes for Down syndrome has reignited this debate about pressure on people to accept screening and testing for this condition during pregnancy. As the use of this non-invasive prenatal test does not involve the risk of miscarriage and removes the need for an invasive test, pregnant people may find it more difficult to refuse as they cannot cite the danger of miscarriage as a reason for refusal. A recent Canadian study considered whether routinization of and public funding of non-invasive prenatal testing for Down syndrome would cause pregnant people to be concerned about pressure to accept the test and to opt for termination of pregnancy. Of the 882 pregnant people surveyed '63.9% said they would personally feel no pressure to use NIPT [non-invasive prenatal testing] due to routinization and 2.8% said they would feel a lot of pressure [...] 38.7% of pregnant women were not concerned regarding routinization causing increased pressure to terminate in the case of a Down syndrome diagnosis.' As a result, this study shows that a large minority of respondents (36.1 per cent) were concerned about pressure to accept routinely offered non-invasive prenatal testing, and a significant majority of respondents (61.3 per cent) were concerned about pressure to terminate pregnancy as a result of the routine implementation of non-invasive prenatal testing.[54]

Routine screening programmes for Down syndrome are usually justified on the basis of empowering pregnant people with the information they need to make an informed choice about their pregnancy. As a result, it is usually recognized that enabling the best standards of informed consent, including providing balanced information and non-directive counselling on Down syndrome, is important when it comes to this type of routine screening.

However, it can be argued that the recommendation that is implicit in the routinization of screening and testing is hard to reconcile with a balanced view of what it means to live with Down syndrome. While attempts are made to remove value-laden language such as 'abnormality' and 'risk' in favour of more neutral terms such as 'condition' and 'chance',[55] the fact that a routine screening programme exists for Down syndrome sends an unequivocally negative message about the condition. As one woman put it, 'Because you have a test [during pregnancy] you think that it must be a terrible thing if it [a diagnosis of Down syndrome in the foetus] happens.'[56] This negative message is one that may undermine any requirement to ensure that consent is well informed and free from influence to choose a particular way.

Understanding the background of routine screening for Down syndrome can help us to understand why there often seems to be an incompatibility between routine prenatal screening and gaining voluntary informed consent. There is strong evidence to suggest that the introduction of routine prenatal screening was motivated not by women's empowerment (and the empowerment of other pregnant people) but by what are called public health goals.[57] The public health goals in this context are to reduce the incidence of certain conditions in the population and, in the case of untreatable conditions such as Down syndrome, by preventing the birth of those with this condition, although this rationale 'has been mostly unspoken'.[58]

The development of routine screening for Down syndrome was at least in part motivated by concerns about avoiding the financial burden that those with this condition were thought to present. The cost of providing amniocentesis to pregnant people at higher risk of having a child with Down syndrome was routinely compared with the costs of care for those with Down syndrome.[59] In addition to concerns about the costs of allowing children to be born with conditions like Down syndrome, routine prenatal screening for Down syndrome has been motivated by a concern about litigation from doctors where parents are not aware of their child's diagnosis before birth,[60] financial considerations (prenatal screening has huge commercial value[61]) and by notions of the welfare of future children born with what are usually considered to be disabling and undesirable conditions.[62] Understanding the historical origins and motivations of prenatal screening helps us to understand why a high uptake of screening is seen as desirable and why information and policies around screening may not emphasize choices to refuse screening or to continue with the pregnancy after a positive diagnosis.[63]

There is evidence that information pamphlets are biased in favour of screening[64] and that while there is good practice in terms of offering balanced information in some cases, the quality of the information given depends upon the particular healthcare professional who provides this information and how they view this offer of screening.[65]

Further, guidance for healthcare professionals may reinforce these public health goals of routine screening. For instance, a report into the ethical implementation of non-invasive prenatal testing into existing routine screening programmes in the UK in 2017 argued that the Royal College of Obstetricians and Gynaecologists guidance *Termination of Pregnancy for Fetal Abnormality in England, Scotland and Wales* 'should be renamed immediately to indicate that they cover the continuation of pregnancy after a diagnosis of fetal anomaly, and this part of the guidance should be expanded significantly, or additional guidelines created. […] In addition, the National Institute for Health and Care Excellence (NICE) should produce clinical pathway guidance on the continuation of pregnancy after diagnosis of fetal anomaly'.[66]

## What justifies routine prenatal screening and why is this important?

There are a number of different things that could be motivating the routine nature of prenatal screening for Down syndrome. As we have seen, this screening is usually justified as a way of empowering pregnant people with information that might enable more informed choices about their pregnancies and their lives. However, it is difficult to accept this as either the main or sole goal of routine screening for a number of reasons.

First, as we have seen, it seems that the routine nature of this kind of screening and the recommendation this seems to imply may be incompatible with gaining voluntary informed consent.[67] Second, it may be argued that there is a better route to empower pregnant people regarding Down syndrome. It has been suggested that this might be better achieved by routinely offering balanced information about screening and testing, rather than directly and routinely offering the screening and testing itself and allowing patients to opt into screening and testing if they decide to do so.[68] This would fit with our usual approach to genetic testing in the non-pregnant population. In this way, it may be possible to better inform pregnant people about this condition and the possibility of screening and testing for this condition without the challenges to voluntariness that are presented by the routineness of screening.

Further, even if we could make a case that routine screening for Down syndrome was justifiable as it empowered pregnant people with information important for making informed choices about their pregnancies and their lives, then it is not clear why this should only be a good idea in the case of Down syndrome. There are many other conditions that might empower pregnant people's choices here, including adult-onset conditions such as Huntington disease. There are calls to massively extend routine screening in pregnancy by extending the use of non-invasive prenatal testing and whole genome sequencing.[69] The routinization of Down screening has been accepted as 'just another blood test' as part of prenatal care despite the complexity of the issues it raises, including issues around gaining voluntary informed consent. I argue that it is time to evaluate the ethical appropriateness of current routine screening before this model of routine screening is used as the basis for extending routine screening using non-invasive prenatal testing and whole genome sequencing for myriad conditions.

Justifying routine prenatal screening for Down syndrome based on public health goals is challenging. Down syndrome has been a focus of prenatal screening from early in the development of these programmes. The rationale behind this focus has often been the financial and other burden that individuals with this condition place on a society.[70] However, in the case of Down syndrome, justifying this approach and considering individuals with this condition as a 'burden' is simply inaccurate in many cases. While some individuals born with Down syndrome may need more publicly funded medical and educational support than other individuals, this is not always the case, particularly with the development of more opportunities for those living with this condition.[71] Further, even if we accept that there is some level of financial burden involved with the birth of some individuals with this condition, it does not seem to be a robust reason to justify routine screening that inevitably seems to involve some level of coercion for some pregnant people. If we accept that screening is justified to save money, then it seems that we could justify all sorts of other interventions based on the same premise – for instance, limiting the number of children that individuals are supported to have. It has also been argued that attempting to reduce the costs of care for those with Down syndrome through the existence of routine prenatal screening 'may inadvertently increase these costs by propagating attitudes which restrict these people's independence and employment opportunities'.[72] The establishment of routine screening seems likely to perpetuate the often overly negative attitudes we may have to those with Down syndrome, which may then have an impact on the opportunities available to those with this condition.

The only other possible justification for the routinization of prenatal screening is evidence of harm prevention. It may well be that we are justified in recommending and even attempting to influence individuals' choices if there is good evidence that doing so will prevent serious harm to them or to others. This is, of course, the rationale behind government screening programmes for breast and bowel cancer. More extreme examples of justified overriding of individual autonomy to prevent harm might also include compulsory treatment and autopsies, speed limits and pressure to accept vaccinations.

When I have argued that we should move away from routinization of prenatal screening for Down syndrome (perhaps moving to policy of voluntary choice or opting in rather than routinized screening), it is this notion of prevention of harm, either harm to the pregnant person or the future child, that is usually invoked to oppose this move.[73] It may be that a pregnant person wants to have this information about Down syndrome and not having it is harmful to them and their decision making, but in this case, an offer of opt-in testing for this condition would allow them to make this decision for themself in a way that was much more in line with the usual way that we choose diagnostic tests. A routine offer of an elective test along these lines would arguably be more empowering for pregnant people's choices and be more likely to remove the value-laden element of routine screening programmes.[74]

While pregnant people could gain access to testing and screening for Down syndrome without routinization of screening, the argument might be that by moving away from routinization we might allow more individuals with Down syndrome to be born and that this would be a bad thing in terms of the welfare of the children they would become. Whether or not this argument stands up to scrutiny is something we will explore in detail in the coming chapters. However, it does seem that the only *seemingly* palatable reason for routinization is the notion of the welfare of the future child, and this is the reason that I include this slightly different example in the three examples I focus on in this book.


## The huge appeal of regulation based on the notion of the welfare of the child

Through the examples I have explored above, which illustrate the three areas of regulation that I focus on in this book, I have tried to show the significant extent to which regulation around reproduction focuses on notions of the welfare of future children. This is the case not just with regulation that explicitly mentions the welfare of future children like that governing access to fertility treatment, but also the implicit justification of routine prenatal screening for conditions like Down syndrome. These regulations are common throughout the world. The use of IVF is increasing and it is expected that we will see an expansion of both pre-implantation screening of embryos and prenatal screening in the near future with the addition of whole genome sequencing techniques. It is likely that the established blueprint for regulation, with its central focus on the welfare of future children, will be used as the basis for future regulation in this area. This is particularly likely given that for many of us our intuition tells us that this prioritizing of the welfare of future children is the right thing to do when it comes to regulating reproduction.

It is similarly clear to many of us that the notion of respect for individual autonomy is another central ethical principle and one that is usually highly prized in modern democracies. The principle of respect for individual autonomy is the idea that we should respect the choices of individuals about how they live their lives wherever possible and work to ensure that these choices are as authentic as possible by removing obstacles to effective decision-making, such as coercion or lack of sufficient information. It is this notion of respect for individual autonomy that is the foundation for our requirement for voluntary informed consent for any medical procedures. However, while respect for individual autonomy is seen as important to the flourishing of individuals, in order to respect this principle for all, there must be limits. Thus, to protect others from harm, it can be justified to override individuals' choices in some circumstances. This is, of course, the rationale behind public health measures like enforced quarantine and other restrictions such as speed restrictions on our roads.

At first glance, regulation based on the welfare of future children seems to fit within this model well. In line with the notion of respect for individual autonomy, we usually respect and, as far as possible, enable individuals' choices about whether and how to have children, and this position is enshrined in law and protected by legislation, including Article 16 of the Universal Declaration of Human Rights and Article 12 of the European Convention on Human Rights. However, this does not mean that all reproductive choices must be respected as some reproductive choices may cause unacceptable harm to others. This view is put forward by the *Brazier Report*, which argues that the right to reproductive or procreative autonomy is not 'an absolute right, especially since it can come into conflict with the rights of others. Procreation is not just a matter of individual freedom. It entails bringing about the life of another human, whose welfare and autonomy deserve the highest attention from the state'.[75] It is this sense of weighing the respecting of reproductive choices of the individual against considerations of the welfare of children who will be brought to birth as a result of these choices that is the basis for the regulation in this area.


## Justified harm prevention?

The assumption that these kinds of regulations, focused on the welfare of future children, are effective and necessary to prevent serious harm to future children is one that has a great deal of support. As we move on through this book, I hope to explain why we should question this support, not because we shouldn't be concerned about the welfare of future people, but because regulations developed on this basis are not, I argue, effective in preventing harm to individuals in the way they are intended to and may even cause harm. But before I explore these substantial limitations of this regulation, it is important to recognize the significant support that these regulations have in order to understand why questioning their appropriateness is such a difficult but, for me, such an important thing to do.

The regulations we are focusing on in this book are based on the idea that where choice is possible, we should avoid bringing to birth children who will experience challenging conditions either physically or socially. In the case of access to fertility treatment, we feel a responsibility to avoid creating children in circumstances that we consider will not provide them with a reasonable quality of life. When it comes to pre-implantation genetic testing and the selection of embryos, we may feel that it is unacceptable to choose to implant an embryo with a condition like deafness when there are alternative embryos which do not seem to have any conditions that we consider may reduce their quality of life. We might accept a degree of pressure when it comes to routine screening for Down syndrome, partly because we subscribe to the idea that trying to ensure that children are born without conditions that many of us see as negative is a legitimate aim when it comes to prenatal care.

This concern is understandable and widespread. Our instincts to protect future children mean that many of us react to choices to bring to birth a child with a condition we feel will impact negatively on that child with concern and even outrage.[76] This was seen in the response to the 2002 case of a lesbian and deaf couple from Maryland who set out to have a deaf child by intentionally selecting a deaf sperm donor. The case caused a huge amount of controversy with commentators arguing that 'parents have violated the sacred duty of parenthood, which is to maximise to some reasonable degree the advantages available to their children',[77] that '[t]his couple has effectively decided that their desire to have a deaf child is of more concern to them than is the burden they are placing on their son'[78] and that this choice was 'incredibly selfish'.[79]

## Academic support

There is a great deal of academic support from highly influential academic ethicists for the idea that we may have a moral obligation, where choice is possible, to choose to bring to birth the 'best' child we can, that is, to avoid disability or the creation of 'suboptimal' lives wherever possible and that to do otherwise is morally unacceptable.[80] For example, John Harris argues that 'it may be morally wrong to "choose" to bring to birth an individual with any impairment, however slight, if a healthy individual could be brought to birth instead'.[81] Similarly, Julian Savulescu and Guy Kahane, argue that 'it is in fact implicit in commonsense morality that it is morally permissible and often expected of parents to take the means to select future children with greater potential for well-being'.[82] Jonathan Glover voices similar sentiments when he writes, 'Consider the theoretical possibility of screening to ensure that only a disabled child would be conceived. This would surely be monstrous. And we think it would be monstrous because we do not believe it is just as good to be born with a disability'.[83]

I have spent my career working to show why these academic arguments, while initially intuitively appealing, do not stand up to scrutiny and we will explore these arguments as this book continues. However, for now, the point I am making here is that these arguments, that we have a moral obligation to bring to birth the 'best' child possible, are put forward by high-profile ethicists who have a great deal of influence, and these arguments lend substantial support to the sort of regulation we are concerned with here, based on concern for future children.

## Healthcare professionals

As well as fitting with many of our intuitions around child welfare and with high-profile academic arguments, this idea that we have an obligation to bring to birth the 'best' child possible fits with the views of many healthcare professionals. For instance, those working in the area of assisted reproduction are generally very supportive of the use of the Welfare of the Child provision to regulate this access.[84] Healthcare professionals working in this field often feel that the use of this assessment fits with their 'heightened sense of professional accountability' when it comes to the role they play in enabling these children to be born. This sense of responsibility is expressed well by one participant in a study around this issue who said, 'It's like they wouldn't exist without you – so you're involved in the existence of something, so you have to be responsible for making sure that that existence is as nice an existence as possible.'[85]

This sense of responsibility for the welfare of children who result from treatment is also expressed by healthcare professionals involved in the use of pre-implantation genetic testing. For example, in response to the proposed amendments to the UK law on the use of pre-implantation genetic testing in 2008, Peter Braude, at the time director of the country's leading centre for pre-implantation genetic testing at Guy's and St Thomas' Hospital in London, reportedly said, 'I have serious concerns about deliberately selecting an embryo for deafness. This is the same as taking a normal child and deliberately making it deaf so that it can fit in with a community. I don't see how that can be acceptable.'[86] Similarly, Allan Templeton, head of obstetrics and gynaecology at Aberdeen University said, 'there is a very real conflict between parental desires and the welfare of the child'.[87]

When it comes to routine screening for Down syndrome, the existence of such screening programmes receives a great deal of support from healthcare professionals.[88] There is also evidence that most people would prefer not to have a child with Down syndrome based on all sorts of reasons, including reasons to do with their own financial and psychological wellbeing but also to do with concerns about the welfare of a child who has this condition.[89] It is unsurprising then if healthcare professionals involved in routine prenatal screening often share these general attitudes.

## Is this the full picture?

There appears to be widely held support for the notion that we have an obligation, where choice is possible, to choose to bring to birth the 'best' child possible and widely held support for regulation that reflects these views and requires consideration of the welfare of future children when making reproductive choices. As such, this seems to indicate that there is a general consensus on allowing some kind of influence to be used in such cases to attempt to prevent the creation of these future children about whose welfare we are concerned. But is this the case and, even if it is, does that mean that we should continue to accept these regulations unquestioningly?

While this kind of response may be the prevailing view on these issues and the stance that is generally taken in regulation, this is not the full picture here. On the other side of this debate, there are those who argue that these regulations cannot be justified and risk causing harm to individuals. For instance, there may be individuals who feel that they were coerced into accepting routine screening for Down syndrome and, in some cases, even felt pressured or coerced into a termination as a result of Down syndrome being identified in a foetus during pregnancy. There are those who feel that making screening for Down syndrome a routine part of prenatal care sends an unwarranted negative message about the condition they or their loved ones live with. There are individuals who, like many other prospective parents, wish to have a child who is 'like them' and so might wish to have a child with the same disability they have like deafness or achondroplasia. There may be individuals with learning disabilities or autism, or those who face other challenges, who are attempting to access fertility treatment and finding getting the same access to treatment highly problematic when they simply want to parent a child as they would have been able to do if they happened to be naturally fertile.

I have found the general position on the issues that underpin these regulations problematic for years and have spent a great deal of my academic career questioning whether regulations that aim to influence the reproductive choices of individuals on grounds of protecting the welfare of the child are as justifiable as they may first seem. As we have seen, in a society where respecting the individual values and choices of others is usually seen as something we value highly, attempting to influence these choices can only be justified by a strong argument that doing so prevents serious harm. For me, this strong argument in favour of these practices is simply not there in most of these kinds of cases.

In the rest of this book, I will explore these issues in more depth to enable you to strengthen your own conclusions on these issues. I will argue that, at least in most situations where we are concerned about the welfare of future children, this does not give us a strong justification for attempting to influence or interfere with the reproductive choices of others. However, the aim of this book is not necessarily to get you to agree with me. The aim of this book is to enable you to examine the positions you hold on these difficult questions and encourage you to re-evaluate the central issues and counterarguments in this debate, to ensure that the position you take is one that you are confident in defending against possible counterarguments. This will involve examining where you stand on ethical questions around disability and screening, and considering different viewpoints to your own and being prepared, if necessary, to change your mind (something we all find very difficult!). In the next chapter, I suggest a way to approach these ethical questions and why taking this open and questioning approach is essential when it comes to tackling complex ethical questions.

# CHAPTER 2
## HOW CAN I DEAL WITH ETHICAL QUESTIONS WITH CONFIDENCE?

## Introduction

In the previous chapter, we considered the widespread regulation in the area of reproduction based on concern for the future child. I suggested that even though these regulations have a great deal of support and seem to fit with many of our shared intuitions, if we want to ensure we arrive at a position on ethical questions we can be confident of, we must consider the other side of the argument and be prepared, if necessary, to change our minds.

In this chapter, we will explore, in a bit more detail, why, if we want to arrive at a robust position on an ethical question, we must take a questioning approach rather than simply accept current regulation or rely on our own intuitive answer or even public opinion. I suggest a simple 'toolkit', or practical approach to dealing with ethical questions, like the ones we focus on in this book. Those of you who already have experience in dealing with ethical questions might still find my suggested approach in this chapter interesting and useful, but otherwise feel free to move to the next chapter where we start to explore the questions and arguments underlying these regulations in more detail.

## How can I deal with ethical questions with confidence?

If you ask people *'What is ethics?'* you are likely to get a number of different kinds of replies. Most will agree that ethics has to do with the question of how we should behave and what choices we should make about how we live our lives. As a result, ethical questions are usually questions about whether we should do something or allow something to be done.

Medical ethics, healthcare ethics and bioethics consider all sorts of ethical questions in the areas of medicine, healthcare and the biosciences, respectively. The classical sorts of questions focus on issues of abortion, assisted dying, resource allocation, confidentiality, the limits of respect for autonomy and questions around the moral status of embryos and foetuses. Ethical questions in this area of inquiry are typically controversial, often with no clear consensus as to what should or should not be done in relation to them. As such, the questions we are focusing on in this book are not unusual in this field in that there is often no clear right or wrong answer to these questions as we can recognize merit in arguments on different sides of the debate around these questions. Earlier, I asked you to consider the questions we will be focusing on in this book and make a note of your thoughts about these questions:

- *Question 1: Are we justified in attempting to evaluate the potential parenting ability of those trying to access fertility treatment (e.g., disabled people or individuals with past criminal convictions) and prevent access in some cases?*

- *Question 2: Should we allow prospective parents using IVF to implant an embryo with a condition considered a disability? For example, should a deaf person be allowed to implant a 'deaf' embryo?*

- *Question 3: Is routine screening for Down syndrome in pregnancy ethically acceptable even if there is evidence that individuals may feel pressure to accept this screening?*

If you did think about these questions and noted down these thoughts, how did you find this experience? Was it straightforward? Perhaps you were very clear from the moment you were asked what your answer would be. Did you find answering these questions challenging? What process did you use in answering these questions, for instance, did you think about the possible counterarguments to your position before arriving at your answer?

In my experience, while there will have been some of you who found answering these questions very straightforward, the majority of people will find answering these questions quite daunting. Without a clear consensus, there are no easy or clear answers to these complex ethical questions. Whatever position we take, there will be those who do not agree with this position. Views on these questions are often polarized, and as a result, it is often impossible to reach a true compromise in the sense of meeting in the middle or splitting the difference, and thus, we are left with trying to find another way forward.

## Why not just focus on current regulations and leave it at that?

We could answer our three questions by simply stating what regulation requires us to do. So, to answer our three questions we might say,

- Question 1: Currently, many jurisdictions require that we consider the welfare of the child before providing access to fertility treatment. Where this is the case, healthcare professionals providing fertility services must consider the welfare of the resulting child and prevent access to fertility treatment if they feel that this child's quality of life is likely to be below a reasonable level.

- Question 2: In places like the UK, it is not legally permissible to choose to implant an IVF embryo that has been identified as having a condition that is generally seen as disabling (unless there are no 'unaffected' embryos available for implantation). In other jurisdictions, while it may not be explicitly prohibited to implant an IVF embryo that has been identified with a 'disabling' condition, the use of pre-implantation genetic testing is usually confined to selecting against 'disabling' conditions, and thus, using this procedure to do otherwise is unlikely to be accepted.

- Question 3: While studies consistently show evidence that routine screening for Down syndrome may involve some pressure on pregnant people to accept screening, this type of screening is a well-established and accepted part of prenatal care internationally.

Current regulations tell us what we are allowed or required to do at the moment in our own jurisdiction. Knowing what these regulations are is important, and we should, of course, always act within these regulations to avoid prosecution or accusations of malpractice. However, for those of you who are professionally or personally involved in this area, it is also imperative to understand and address the ethical questions here. Instead of asking what we are *allowed* or *required* by law or policy, ethical questions ask us to explore what *should* be allowed or what *should* be accepted practice in these areas of reproduction. Exploring the question of what our regulations *should* be in these areas of reproduction will allow you to come to a position on whether our current regulations can be ethically justified. This will involve evaluating the ethical concepts and arguments that underlie current regulation, and doing so is important for the following reasons:

1. While it is imperative that we always work within our local and current regulations, as we can see from our three questions, regulations often leave a great deal of room for interpretation and do not provide the 'black and white' guidance that we might wish to have. A good example of this is the Welfare of the Child requirement of the UK HFE Act.

This requires that clinics must consider the welfare of the child and 'decide whether there is a risk of significant harm or neglect to any child[1]' before allowing individuals access to treatment. However, the regulation here does not indicate what might be meant by 'significant harm' or give any further indication as to when treatment might be refused. It is up to the clinics and those involved in these decisions to weigh up the evidence and make a judgement here. An understanding of the often-complex issues, arguments and questions that underlie this assessment will be fundamental in the process of reaching decisions that we are able to defend.

2. Again, while it is always important to work within the current regulations, it is important to re-evaluate established regulations to ensure that these regulations do the job they were developed to do and, in these cases, that any infringement on individuals' choices can be justified. Just because regulations are well established and are enacted around the world doesn't mean that we should simply accept these. Overturning regulations that criminalized homosexuality and denied equal treatment of women and people of colour are strong examples of why it is important to continue to question established regulation. When it comes to our three questions around reproduction, it might be that you are part of a pressure group, or you are an individual attempting to change the way you are being treated or someone who is responsible for reviewing current regulation in this area. Gaining a strong understanding of the ethical arguments that underpin these regulations is not always an easy task but one that will be important if you want to either be confident that the existing regulations are fit for purpose or if you want to try and convince others that we should make changes to these regulations that have become an accepted part of our regulatory framework.

## How do we answer the ethical question, 'What should we do?', with confidence?

When you consider ethical questions like the ones I have introduced in this book, you may well have been able to give answers, to at least some of these questions, with some conviction and even confidence. We all make complex judgements and decisions every day and we learn to rely on our initial intuitions to make these decisions.

We all have strong feelings about all sorts of issues, and often, these are particularly strong when they involve ethical questions and questions relating to reproduction. However, I argue that we need to do more than give our initial, intuitive responses when it comes to answering ethical questions, particularly when our answers to ethical questions might impact on others' choices.

I talk a lot about intuition in this book, so it makes sense for me to be clear about what I mean by this often difficult to pin down concept.

## The role of intuition

Saying what we mean by intuition is not as straightforward as we might think. Definitions of intuition vary immensely over time, between people and between disciplines. For instance, in psychology, thinking and decision-making is typically seen as a 'two-process model', where intuition is seen as belonging to the more primitive and irrational mode of these two processes of thinking. The psychological concept of intuition is, therefore, associated with fast decision-making undertaken by the unconscious mind. It is contrasted with the other mode of thinking undertaken by the conscious mind, which consists of slower, more deliberate, rational decision-making.[2]

However, there is much less consensus about the role of intuition in philosophy and therefore in ethics, which, as a subset of philosophy, largely shares its methodology. Intuition is a staple part of philosophical argument, with hypothetical thought experiments regularly used to elicit an intuitive response that is often central to an argument. A famous example of this is the thought experiment put forward by Judith Jarvis Thomson that uses the hypothetical and fantastical example of a violinist attached to you for life support but aims to make a point about the ethical justification of abortion.[3]

However, even though it seems intuition does have a role in contemporary philosophy, there is a great deal of disagreement about what this role is and should be. For some, intuition does a good job of giving us the answers we are looking for, whether it is a classical philosophical problem like the nature of personal identity or solving ethical problems. But for others, intuition alone is not a sound basis for decision-making.

Leon Kass argues that repugnance or a negative intuitive response to something should be taken as clear evidence that this something is harmful or ethically problematic.[4] Harris, on the other hand, expresses his concern about the importance Kass places on intuitive response in solving ethical issues:

> The problem that confronts Kass – and anyone who wishes to cut to the chase of morality so to speak by finding ways to decide or to act 'immediately and without argument' – is to have a way of knowing when one's sense of outrage, or ones 'feelings' or whatever, are evidence of something morally disturbing and when they are simply an expression of bare prejudice or simply an induced emotional response.[5]

Further, Harris argues that morality and moral principles must be backed up by reason

> And this means they must always be prepared to offer a reasoned defence and justification of their morality or elements of it. It would never be enough or indeed even respectable for the reply to be 'I just felt like it'.[6]

For those, like Harris, who do not think that intuition alone can provide us with reliable answers, what is needed is deliberation and reason to evaluate and, if necessary, moderate these intuitive responses.

This is the view I take of the role of intuition in ethical deliberation and decision-making. While our intuitive responses might be a reasonable starting point when faced with an ethical question, I argue that we need to rely more heavily on reason, logic and more-developed deliberation in order to arrive at decisions that we can be confident we can justify. This is the thinking behind the approach or toolkit I recommend in this book, which I will explain in more detail in a moment.

## The reality of unconscious bias

The problem of relying on our intuition when answering ethical questions is that while we might be convinced that this is the right approach, the initial answer we give may be one that has arisen from unconscious thought processes. It seems that a great deal of our thought process is unconscious.[7] As Timothy D. Wilson explains 'According to the modern perspective, Freud's view of the unconscious was far too limited. When he said [...] that consciousness was the tip of the mental iceberg, he was short of the mark by quite a bit – it may be more the size of a snowball on top of that iceberg.'[8] As a result, a great deal of our mental processes, including judgements, feelings and motives,[9] occur in parts our minds that cannot be accessed by our conscious minds.

Unconscious thought processes influence all human interactions, tend to reinforce the 'values' of the society we live in and are very often difficult to identify and correct[10]. One aspect of our unconscious thought processes is what has become known as unconscious bias. Unconscious bias takes all sorts of forms. Confirmation bias, for instance, is 'a type of unconscious bias that causes people to pay more attention to information that confirms their existing belief system and disregard that which is contradictory.'[11] Affinity bias is the predisposition to seek out and view favourably people who are most like ourselves.[12] Beauty bias is the tendency to prefer individuals who are better looking when it comes to recruitment.[13] Conformity bias describes how peer pressure is likely to influence us to conform to the opinion of others around us.[14]

It is likely that we are all affected by countless different forms of unconscious bias in many of the choices or opinions we make every day.[15] But given that this bias is unconscious, it will be something that most of us do not realize affects our decision-making. Anyone who has completed a training course on unconscious bias or completed an online unconscious bias test[16] will probably have been surprised as to the extent to which their own judgements are affected by bias.

The possible influence of these unconscious factors may mean you might not be able to defend the conclusions you come to if called to do so. Due to the skewing effect of unconscious bias, your answer may also not represent what you might, on more reflection, consider to be your core values. For instance, individuals who pride themselves on believing in equality are just as likely as anyone else to display implicit bias around sex, race and disability.[17]

There are things we can do to try and mitigate the influence of unconscious bias and allow us to come to a position that aligns to our core values and that we are able to defend. Most institutions, for instance, now integrate procedures to mitigate unconscious bias in processes such as recruitment or assessment.[18] This might include a focus on meeting certain criteria in order to be called to interview for a job, or ensuring that formal assessment, where possible, is done anonymously in order to prevent unconscious bias affecting the outcome of the assessment. In the same way as these procedures aim to mitigate unconscious bias in intuitions, I suggest that following what I call the ARC approach will help us to mitigate the influence of these unconscious biases and develop a reasoned and nuanced position when it comes to answering ethical questions.

## The ARC approach – a toolkit for dealing with ethical questions

To try and guard against the influence of unconscious factors such as bias and to provide a response to ethical questions that – you can be more confident – reflects your conscious thinking, I suggest that you use what I call the ARC approach. This ARC approach represents a commonly used approach when it comes to ethical questions but is expressed in a way that I hope is clear and easy to remember and to apply.

This approach suggests that first, you need to identify the ethical question you wish to address and make sure it really is an ethical question rather than a legal or clinical question. An ethical question will be a question about what we *should* do or what is *ethically justifiable* rather than what is legally permitted or what the accepted clinical approach is to an issue. Once you have identified your ethical question, I suggest that if you want to arrive at an answer you can be confident about holding, as well as thinking about your answer to your ethical question (A) you should consider two further questions (R and C):

- **A:** What is your ANSWER to this ethical question?
- **R:** What is your REASON for your answer?
- **C:** What is the main COUNTERARGUMENT to your position, and how would you deal with this counterargument to defend the position you take?

By following this approach, we do more than just provide our answer to any ethical question, an answer that might rely on intuition and unconscious thought processes. If we answer the two subsequent questions, we can try to guard against the influence of skewing factors such as bias and to provide a more nuanced and reasoned response to ethical questions.

Being explicit about your reason for the answer that you give will start to develop your position in more detail. The reasons you give for the position you take are likely to involve ethical principles such as respect for autonomy and avoidance of harm or fairness. These reasons may draw on the arguments of others – in fact, reading around these ethical issues is a really good way of developing your own position on ethical questions.

By considering the arguments on the other side of this debate seriously, we can test the position we have taken to see if these counterarguments illuminate any weaknesses in the position we have taken. If we can show how our position stands up to these counterarguments we are able to defend our position and make it stronger. By ensuring you have answers to these three questions about the position you take, you will have ensured that you have provided an answer that is much more than an intuitive response to this question. You will have provided an answer that you can and do defend and show reasons for.

This approach involves gaining an understanding of the issues around the ethical question you wish to answer and then coming to a position that is more than just a statement of what you *feel* is the right thing to do but is, rather, a position that you have really examined in detail and provided reasons for and can defend against the main counterarguments that might be levelled at it.

You might find that in the process of trying to answer the questions that make up the ARC approach you do not have clear reasons for your answer or you find it difficult to defend your answer against possible counterarguments. This may mean that you need to do some more work on exploring your reasons for taking this stance to get to the point that you can defend it. It may mean that on reflection you might consider changing your stance if you feel that while you have a strong intuition about this issue, you cannot, thus far, defend your position in this way. This can be an uncomfortable situation to be in. However, without being able to identify reasons for taking a particular position and the ability to defend that position against counterarguments, your position is very vulnerable to criticism. As we have seen, I suggest that we need more than our intuitive response if we are to arrive at a robust position that we are able to hold with confidence, and the ARC approach provides a way of developing this position.

Of course, there will be some of you who disagree with this approach and argue that focusing on intuition does provide a good way of guiding our answers to these ethical questions. For those who retain this view, I suggest they consider some instances where this focus on intuitive responses seems to have failed to identify strong ethical positions. Take, for example, the criminalization of homosexuality. In the UK, for example, prior to 1967, homosexual activity even in private was a serious criminal offence that resulted in many prosecutions, incarcerations, chemical castration and loss of life.[19] Understanding what motivated criminalization of homosexuality is difficult for many of us from our modern perspective, where attitudes, at least among younger generations, have generally become much more positive. It can seem difficult in hindsight to understand how criminalization and stigmatization of this section of society was sanctioned by governments for so long. When we explore the motivation for the persecution of homosexuality, we find that intuitive responses, often in the form of disgust, are a main motivation.[20] This intuitively negative reaction was often justified with ideas of harm prevention, with a common belief that homosexual individuals may 'corrupt' or 'recruit' others, particularly the young.[21] While there is no doubt that those who supported criminalization of homosexuality did so on the basis of strongly held views that allowing this behaviour to go unchecked was unacceptable, hindsight provides clear evidence that the criminalization of homosexuality cannot be justified by the use of offence and harm prevention. It seems that the sort of unconscious bias that reaffirms cultural norms may explain why it took so long for attitudes to change and thus for the law in this area to be repealed.

## Conclusion

Answering ethical questions like the ones that we are focusing on in this book can be challenging as there is often no clear consensus as to what is the right thing to do. In this chapter, I have suggested that while our initial answer to an ethical question can be a good starting point, we need to do more than simply provide this initial answer if we want to come to a position on ethical questions that we can be confident of holding. Our initial answer may be influenced by all sorts of unconscious thinking, including unconscious bias. To try and mitigate the effect of unconscious bias and to develop a detailed, nuanced and robust answer to ethical questions I have suggested you use what I have called the ARC approach. This approach encourages you to do more than just provide your **Answer** to the ethical question. It encourages you to go further and provide your **Reasons** for your answer and consider the **Counterargument** to your position and show how your position can stand up to this counterargument. This is, of course, not the only way of responding to ethical questions but is one, I suggest, that will enable you to develop ethical positions that you can defend if you need to and, thus, that you can hold with more confidence.

## Try the ARC approach

If you did make a note of your answers to our three questions at the start of the book. Go back to these now and ask yourself not only:

**A:** What is your ANSWER to this ethical question?

But also:

**R:** What is your REASON for your answer?

**C:** What is the main COUNTERARGUMENT to your position, and how would you deal with this counterargument to defend the position you take?

Do not worry if you still find this difficult; the rest of this book will help you to pick apart the arguments on both sides of these questions, so that you can answer these questions, using this ARC approach, and arrive at a position you can be confident in defending.

# CHAPTER 3
## WHEN IS HAVING A CHILD
## A HARMFUL THING TO DO?

## Introduction

While upholding and enabling the reproductive choices of individuals is usually seen as important to allow individuals to control their own lives and respect the plurality of beliefs and values in our society, we also recognize that there should be a limit to this respect for individual choices, particularly where these choices are likely to cause serious harm to others, in this case, a resulting child. Interfering with individual choices seems justified if, by doing so, we prevent serious harm to others.

For many of us, it may seem obvious that choosing to create a child with a condition such as deafness or where their parents face other challenges, including significant disability or other social issues, is to cause a child to exist with serious challenges. As a result, we might feel obliged to try and influence these choices to prevent the birth of this child, and in doing so, we might argue, we avoid this serious harm.

But, as we have seen in the previous chapter, we need to ensure that the responses we have to these ethical questions around the welfare of future children are more than intuitive responses. Intuition can indicate a strong ethical position, but it can equally result from bias or other unconscious influences. Regulations in this area have the potential to influence and even override individual reproductive choices. Basing these regulations on unexamined intuitions is unacceptable, and the onus is on us to come to a position that we have deliberated more deeply and can provide reasons for and defend against counterarguments. Without this level of justification, we are in danger of overriding the choices of some individuals without good reason.

In this chapter, I return to the question I raised in the introduction to this book 'Is being alive generally a good thing for most people?'. I argue that where we stand on this question will have a strong influence on when it might be considered harmful to have a child, and thus, finding our position on this question will help us to develop our answers to the three questions that are the focus of this book. We often have strong, protective, intuitive reactions to many cases where the welfare of future children is concerned. However, if we think that being alive is generally a good thing for most people, then it is difficult to provide a convincing account of the harm that we are concerned about when we consider choices to bring to birth a child with a disability like deafness or Down syndrome or into a family situation that we might view as suboptimal.

## Is being alive generally a good thing for most people?

In the introduction to this book, I asked you whether you thought being alive is generally a good thing for most people. You may have been surprised at this question as we live in a society where it is usually assumed that the answer to this question is positive. Most of us, perhaps until we are asked this question, assume the pervading cultural norm in our society – that we are lucky to be alive and that having children is a good thing – is unequivocally true. I know that until I engaged with this question myself, I hadn't taken the arguments on the other side of this debate seriously.

Our society is one that might be generally called a *pronatalist* culture, that is, one where the bringing to birth of new human lives is seen as a positive thing, something to be celebrated and even encouraged. This general cultural approach is reflected in many aspects of our lives. For instance, human procreation is usually seen as something important, good and necessary, and for those who are religious, having children may even be seen as God's will.[1] We often share intuitions that life generally is a good thing for us and for those around us, including our children. When asked, we know that most people say that they are generally happy,[2] and from this, we tend to assume that being brought to birth was a good thing for us and that we have benefitted from being brought to birth. Those who disagree are often treated with suspicion or assumptions that they may be mentally ill.[3] Our shared intuitions that we benefit from being brought to existence and that bringing others into existence is a good thing are ingrained in our human culture.

This positive attitude to procreation translates into policies that encourage and support those who choose to have children, for instance, by providing government-funded parental leave and other financial support for those who choose to raise children. Having children is often seen as altruistic, with a feeling that it is virtuous that parents make sacrifices to bring children into the world and care for them. Involuntary childlessness is usually seen as a significant source of sorrow and something to be overcome with fertility treatment or adoption. Voluntary childlessness is often viewed with suspicion and disbelief, and it is often those who choose to remain childless who are called selfish rather than those who have children.[4]

It is easy to see why such strong intuitions are ingrained in humans. Without them, we would have died out very quickly. Our drive to reproduce and our belief that to do so is a good thing has allowed our species to flourish. Most of us share a strong survival instinct so that, despite the challenges of human life, most of us feel that we are lucky to be alive and that we benefit from our continued existence.

## Cases of extremely low welfare

While this shared view that reproduction is a good thing is very widely held and often an integral part of our cultural lives, it is usually qualified by the acceptance that, in some rare cases, human lives aren't such a positive experience. However joyous we are about the idea of new human life, most of us would hold that there are exceptional circumstances where the birth of a new human being is met with sorrow rather than celebration, either because the positive aspects of life are missing due to the lack of consciousness, which might be a consequence of certain extreme disorders such as anencephaly, or as a result of being born with conditions or in conditions that involve extreme suffering that completely outweighs any expected positive experiences.

The lives I have in mind here are those that that are so dominated by suffering that we consider these lives to be a harm overall to those who live them. As we will see later in this book, I argue that these lives are not a harm in a comparative sense that it is worse to be alive in this terrible condition than not to exist. I argue that these kinds of comparisons do not make sense as you cannot compare something (being alive with a terrible life) to nothing (non-existence). But these kinds of lives are seen as a harm in themselves for those who live them as these are lives where the negative aspects of life clearly outweigh the positive aspects, and thus, the experience of living is an overwhelmingly negative one. However, the sort of lives that we would consider to be unbearable in this way are likely to be very rare. For those of us who think that human existence is generally a good thing, these kinds of lives, where we might expect the person living them to welcome death, are very much the exception to the rule.

## *Problems with terminology: 'Lives not worth living' or 'unworthwhile' lives*

When I started writing in this area in the early 1990s, there was a great deal of interest in the academic bioethical literature in this notion of lives that might be considered to be a harm to those who live them. Being able to refer to these lives of overwhelming negative experience and distinguish these from other lives of overall positive value was helpful in discussions of the central ethical questions of assisted dying and screening for disability. These rare lives of overwhelming suffering were often referred to in the academic ethical literature as lives 'not worth living'[5] and contrasted with the majority of human lives that were assumed to be 'worth living'[6] or 'worth-while'.[7] I found this distinction useful, and I too started using these established ways of distinguishing between lives that are considered positive or negative experiences to those who live them, even using the term 'unworthwhile'[8] to indicate these rare instances of lives that were thought to be a harm to those who experienced them.

However, while making this distinction is helpful, I find the use of these particular terms problematic. The way that I and many others have used these terms focus on the value that the person experiencing the life in question may put on their own life. However, the terms 'life not worth living' or 'unworthwhile life' could very easily be interpreted as implying that the lives of others are *externally* judged as worthless and that those living them should be given less respect than others. While we might be legitimately concerned with the welfare of future people and whether they are likely to assess their own life as a positive or a negative experience, any external judgements around the worth or otherwise of another person's life are highly problematic. As soon as we start to judge the quality of another's life not on that person's own welfare or how they will experience their own life but on other factors such as another's individual values or the norms or values of a group or a society, we move into extremely difficult ground.

Outside judgements of the worth of someone else's life has clear eugenic connotations. These connotations of the terms 'life not worth living' and 'unworthwhile lives' are amplified when we recognize the echoes of terminology used in 1930s and 1940s Germany. In August 1939, the Nazis implemented their Aktion T4 euthanasia programme. Under this programme, tens of thousands of individuals with physical or mental disabilities were sterilized and 'euthanised.' While the programme was described as a euthanasia programme, the way that this term was used had very little in common with the modern use of this word. Instead of aiming to provide a gentle or dignified death at the request of individuals in distress, this programme was part of the Nazi commitment to Social Darwinism or eugenics.[9] The term that was used to describe those targeted by the Aktion T4 programme was, in the original German, 'lebensunwertes leben', and the English translation of this is 'lives unworthy of life'; this was contrasted with 'lebeswertes leben' or 'life worth living'.[10] The use of these terms in this context was absolutely one that refers not to value that the person who experiences these lives puts on their life but on the value that is put on lives by others.

The way that I and many others have used the terms 'life not worth living', 'unworthwhile lives' and 'worthwhile lives' in bioethical arguments has absolutely been focused on the value that a person living the particular life in question is likely to put on that life. My work focuses on the welfare of individuals, and, for me, the only value that is relevant here is the welfare of particular individuals, that is, how they experience the life that they live. However, as we will see later in this book, there are positions on the issues we explore in this book that value things outside the welfare of individuals and may be seen as applying external judgements of particular lives.

For the sake of clarity and to emphasize this specific use of these terms in this context, I suggest a change to the terminology here. I propose that instead of 'worthwhile' lives, we talk about 'intrinsically valuable lives' to clarify that our assessment is not an assessment of what we think the value of these lives is in relation to our own or other external value but whether we predict that those living these lives would be likely to value them. Correspondingly, instead of talking about 'a life not worth living' or 'unworthwhile' lives, I propose we use the term 'intrinsically harmful lives'. Of course, attempting to assess how someone else might value their lives is a very difficult thing to do, which we will come onto later. But for now, I hope that this distinction between intrinsically valuable and intrinsically harmful lives is one that can be useful in our discussions to demarcate the threshold between what we might suggest is an acceptable and unacceptable quality of life for future children based solely on our assessment of the welfare of this future child and not on other factors.

This distinction between intrinsically valuable and intrinsically harmful lives can be helpful to express – even if we feel that most lives have positive value to those who experience them – why deliberately creating a life where suffering completely overwhelms any positive value would be seen as a morally questionable choice.

As a result, for those of us who think that being alive is generally a good thing (this is not everyone, but we will come to this later), the answer we might give to the question we started this section with 'When is having a child a harmful thing to do?' would be something like 'Only where the child has an intrinsically harmful life, that is, a life that is completely blighted by suffering so that life is a harm to the individual experiencing that life.' For those who take this view, we can say that it follows from this that there is a strong argument to suggest that where choice is possible, we have a moral obligation to choose *not* to bring to birth individuals who will have lives overwhelmingly dominated by suffering. Cases where we might have this obligation to choose (where choice is possible) not to bring to birth these lives will be rare cases where we predict that the challenges that these individuals will face will be so severe that we predict that life will be an overwhelmingly negative experience for these individuals. Clearly, a judgement about which future lives are likely to fall into the category of intrinsically harmful lives will be a difficult judgement to make. But while this judgement will be a challenge, if you share the view of life that having a human life is usually a good thing but that there are some rare exceptions, this distinction, and the thinking around it, is a useful one when thinking about these cases and the debate they generate.

What we can take from this discussion is that for those of us who think that being alive is generally a good thing but accept the idea that there are some lives that are the exception and are not a good thing for the person living them, then this leads us to the conclusion that while reproduction is acceptable, we do have a moral obligation to avoid bringing to birth such lives of overwhelmingly negative experience that they might be considered to be intrinsically harmful to those who experience them. On this basis, it may be justifiable to enact regulation that aims to enable this obligation and avoid the creation of these kinds of intrinsically harmful lives.

The view that life is generally a good thing but that there are rare exceptions where life is thought to be so bad for someone that it is harmful to them to live that life is a commonly held view and also a good example of the sort of intuitive response we might have to questions about reproduction. Because of our generally pronatalist culture and the widely held views about the good of human reproduction, we often do not question our own answer to the questions I posed earlier: 'Is being alive generally a good thing for

**Figure 1** *The idea of a threshold between intrinsically valuable and intrinsically harmful lives.*

most people?' Unless we are faced with opposing arguments on this issue, we might never question our own position. As I said earlier, exploring the opposing arguments does not necessarily mean you have to change your mind on an issue. However, I suggest that we cannot be confident of our own position unless we consider the other side of the argument, the counterarguments to the position that we take.

When it comes to the question, 'Is being alive generally a good thing for most people?' it is important, then, to consider the main alternative views to the dominant pronatalist view before settling on our answer to this question.

## An alternative view – Antinatalism

Consider this statement reported to have been made by Robert Smith from the band *The Cure* where he says,

> I've never regretted not having children. My mindset in that regard has been constant. I objected to being born, and I refuse to impose life on someone else. Living, it's awful for me. I can't on one hand argue the futility of life and the pointlessness of existence and have a family. It doesn't sit comfortably.[11]

This statement expresses this alternative view to our question about whether being alive is generally a good thing. Smith suggests that since being alive entails suffering that we have not consented to accept, then we should question whether reproduction really is the good thing many of us see it as. This general view of reproduction has been called *Antinatalism*, and those who take this position argue there are no good reasons to reproduce, that refraining from reproducing is the ethically preferable thing to do and that human extinction through lack of reproduction would be a good thing.[12] There are a number of arguments those who take this view put forward to support their position. I will explore these briefly now but recognize that this brief overview cannot reflect the detail of this debate.

## *Antinatalist arguments based on concern for the welfare of the future child*

David Benatar, for instance, argues that 'coming into existence is always a serious harm'.[13] There are several arguments that might be used to support this idea that are based on concern for the welfare of future children.

One argument here, put forward by Matti Häyry, is that it is immoral to have children on the basis that it is 'wrong to bring about avoidable suffering'.[14] He argues that all of us suffer during our lives (pain, fear, frustration and grief, for instance), and thus, by avoiding having children, we avoid causing this unnecessary suffering.

Against this it might be argued that while we will all suffer during our lives, most of us see this suffering as the price to pay for the good things that life allows us, and overall, most of us value our lives greatly.

Those who argue for the wrongness of reproduction might counter this in a number of ways. They might argue, as Benatar does, that we are deluded to think that our lives are good. Benatar argues that 'coming into existence, far from ever constituting a net benefit, always constitutes a net harm.'[15] He suggests that there are psychological reasons why we tend to overestimate our own happiness including tending to remember positive experiences and forget negative experiences. He also points to the many terrible things that happen every day to people including malnutrition, war, rape, assault, child abuse and murder, to illustrate what he feels is the truth about how bad life really is. Benatar claims that 'A charmed life is so rare that for every one such life there are millions of wretched lives'[16] and thus that those of us who believe that we have a life that is intrinsically positive experience are most likely wrong about this.

While Benatar is right that there is a great deal of suffering in the world and that we all inevitably suffer to some extent, accepting his argument that we are deluded in our own assessment of our quality of life is challenging. Most of us value our lives, are happy for our lives to continue and see death as a terrible misfortune even though our lives will contain pain, suffering, frustrations, disappointment and loss. If most people are happy with their lives and feel lucky to be alive, then what sense does it make to say that they are harmed by this experience if this is not their own evaluation of their existence? Given this generally optimistic and positive view of human life the Antinatlist view seems to dismiss unjustifiably the lived experience of most human beings.

In response to this Benatar might argue that even if we accept that we might have good reasons to want to continue our lives, this does not mean that we have good reasons to create new lives.[17] This is based on probably his main argument to support Antinatalism, the so-called Asymmetry argument.[18] The Asymmetry argument refers to the intuitive moral difference between creating good and bad experiences. The argument here is that we cannot have a duty to create happy lives as if we fail in this duty, no one is harmed or made worse off because there will simply be no one to experience this loss. However, it is argued that we do have a duty to refrain from creating lives that either contain bad experiences or are so bad that they are intrinsically harmful. This is because if we do create these lives, *someone* will suffer from these negative experiences. Thus, if we do not have children, no one is deprived of anything. But if we do have children, this will inevitably impose avoidable suffering onto all those we create.

In addition to this argument, it might also be argued that even if we accept that most of us have lives we value, we usually accept that some lives are not good for those who live them – what I suggested could be called intrinsically harmful lives. As a result, it has been suggested that by having children, we are playing Russian roulette with our children's lives,[19] risking imposing lives of overwhelming suffering on the children we choose to have. Future children cannot consent to this risk of overwhelming suffering, and thus, the right thing to do would be to remove the risk of creating these intrinsically harmful lives by not having children.

As well as arguments based on the welfare of the future child, there are other arguments for the Antinatalist view. These refer to the harm and suffering that a continuing human race inflicts on non-human animals[20] and the planet.[21]


*Counterarguments to the Antinatalist position*

There are many other counterarguments to the Antinatlist position that we should explore if we want to come to a position we can be confident about holding regarding the questions on whether human life and human reproduction are generally good things.

One of the most often returned to counterarguments here is the strongly held intuition that we have benefitted from being brought to birth and that we benefit our children by creating them. It is common for people to express how grateful they are that they exist, and perhaps even ponder how lucky we were that our mum met our dad and that the things that needed to come to play for us to be conceived then born happened. Whereas the Asymmetry argument suggests that we cannot have a moral obligation to bring to birth even happy people, our intuitions, perhaps influenced by generally pronatalist social norms, make us feel that it would have been bad if we had not been born. If we think about the possibility of our non-existence, we compare having the life we value to not existing and feel that we have benefitted from being brought to birth and into existence. Sahin Aksoy illustrates this way of thinking when he claims,

> life and existence is always better than non-existence. Therefore, it is irrational and immoral to 'sentence' someone to non-existence while you have the chance to bring them into life and existence. Life may have good and bad days, ups and downs, sufferings and joys – but it is still worth experiencing.[22]

Aksoy is expressing this common-sense idea of comparative harm and benefit here when he talks about the harm of 'sentencing' someone to 'non-existence' and why it is, in his view, 'always better' to come into existence. The problem here is that while we might feel that we have benefitted from a choice to bring us to birth, this comparative notion of harm and benefit just does not work in this context as there is literally nothing to compare with existence.

Thinking about non-existence and particularly our own non-existence is a very tricky thing. A philosophy lecturer once said to me 'I love children. As soon as they are old enough you can tell them that one day they will die and that, up until relatively recently they did not exist. These thoughts will keep them busy for a very long time'. This statement really connected with my own experience of pondering the idea of my own non-existence before my conception and birth. It is a common human phenomenon that we find the thought of our own non-existence, preconception and birth baffling and uncomfortable. This might be why our brains do not seem to be able to accept completely the idea that once, for a very long time, we did not exist and the odds of the person we are (the particular consciousness we experience) existing, was against such extreme odds that we are more likely to win the lottery or get eaten by a shark than exist as the unique genetic person that we are. For me to exist as the person I am involved an infinite number of events to take place in the exact way that they did to ensure that I was conceived in the way that I was, at the moment that I was.

Just as we find it difficult to accept that we might not have ever come into existence, many of us also struggle with the idea of our preconception or pre-birth non-existence as nothing. It is, for instance, very common to give this non-existence state a negative value in comparison to our current life, which most of us generally value and wish to continue. Harris does this when he argues that if a deaf child was brought to birth rather than a different hearing child, that child has 'no complaint because for them the alternative is non-existence'.[23] Here, Harris is expressing a view that many of us have about our own lives that not existing would have been a harm to us and that we feel lucky to be alive and be given the chance to live.

This temptation to compare existence with non-existence also fits with the way the law often deals with so-called 'wrongful life' legal cases. Wrongful life actions are typically brought by or on behalf of a child who claims that because of the negligence of the defendant, they had to endure a life with an impairment or disability. The claim is usually not that the defendant caused the impairment but that but their failure to warn their parents of this impairment caused their parents to continue the pregnancy unaware of this condition and thus cause the birth of this child.[24] These legal cases take place inside a general legal tradition of basing the level of legal damages on attempting to assess the level of harm done to an individual by a particular act. This is usually done by considering what the expected well-being of that individual would be if that act had not taken place – so the extent to which the victim was made 'worse off' by the action. It is not surprising, therefore, against this legal background that those commenting on legal 'wrongful life' cases, have given non-existence a value, usually a value of zero, in order to try and determine the damages that have been incurred when children are born in an impaired state and seek compensation for this impaired existence.[25] If non-existence is given a value, say zero, and the child's life given another value, then an estimate can be attempted as to whether this child is better or worse off by having been brought into existence. As we will explore in detail later, comparative notions of harm and benefit are applicable when comparing possible states of existence. So we might compare the life of someone who is born with or without an impairment that was caused during pregnancy perhaps by the action of a healthcare professional. In this case it would make sense to make comparisons between the possible welfare with the impairment and this person's welfare without the impairment and, if appropriate, attempt to calculate the level of damage to welfare caused. But 'wrongful life' cases do not compare two alternative states of existence but attempt to compare non-existence with existence. Giving non-existence the value of zero in order to try and make the comparative calculations just does not work because when we are attempting to compare existence with non-existence, there is nothing to attach this value to.

The best explanation I have found of why assigning a value, even the value of zero, does not work when it comes to choices and actions that do not change the welfare of a future child, is by David Heyd, who uses the idea of a bank account to demonstrate this.[26] According to this explanation, what those who wish to compare existence with non-existence are trying to do is to say that existence is like having money in your bank account, being in the 'black', having a positive balance to your account because they see having a life as a positive thing, something good. When these people think about not existing, they either see it as a negative or being overdrawn at the bank because they see non-existence as harmful as it removes the chance of existing. Or they see non-existence as neutral, not bad but not good either, and assign it a zero or a bank account with nothing in it, a zero balance. On either of these comparisons, a positive value can be compared against the negative or the neutral and be seen to be better. It is better to have money in the bank than to have nothing or be overdrawn. However, the problem with this comparison is that this gives non-existence or nothing a value even though there is no one and nothing to attach it to. As Heyd puts it, non-existence is not like having nothing in a bank account but more like there being no bank account at all. What we are attempting to compare is something with a positive value (money in the bank) to nothing – not zero but nothing (there is no bank account at all to compare it with).[27] If you do not have a bank account, it does not make sense to say you have nothing in your bank account, as there is nothing to which this value can be ascribed.[28]

Unless we believe in souls somewhere waiting for their chance at life, before a particular individual is brought into existence, there is nothing there, no one to regret a lack of existence, to be frustrated by existing people's choices. We may find it difficult to comprehend the possibility of our own non-existence. I certainly spent many a confused hour as a child trying to imagine what it would have been like if my mum had not met my dad and I was never conceived and thus never existed. But logic tells us, however, that billions of possible lives are not realized every day and we can understand that it makes no sense to regret this on these unrealized individuals behalf as the person these lives would have resulted in has not suffered or lost out as a result.[29] So while many of us are happy to be alive and think ourselves lucky to be here, and thus that life is a good thing, this does not mean that the alternative, never to have been conceived and brought to birth, is conversely a bad thing as it is not a thing at all.

### If we accept the Asymmetry argument, do we have to accept Antinatalism?

Antinatalism is a very interesting position. Many of us will have a strong instinctive reaction to the view that human reproduction might be a bad thing generally speaking. This idea goes against pretty much every social norm and shared values that our societies are built upon. However, when we examine the arguments for Antinatalism, we find that many of them have a lot to commend themselves to us. It's true that none of us have a perfect life without suffering and it also seems true that no sense can be made of the idea that we would experience anything negative if we had not been brought into existence.

But this does not necessarily mean we have to accept the Antinatalist position. Whether or not you accept this position has more to do with your own experience of life and your response to the question 'Do you think that being alive is generally a good thing?'. Our answer to this question is on one level a very personal one but also one that we can ask our fellow human beings. While there is significant variation in how individuals assess the value of their own lives, we know that the majority of people, when asked, assess their lives positively. Further, we know that improving certain social conditions such as levels of government social expenditure are likely to improve this further.[30] While Benatar dismisses these generally positive assessments of our own welfare as delusion,[31] denying the lived experience of many individuals would seem to be something that is difficult to defend and appears to go against the Antinatalist focus on assessing and promoting the welfare of human beings.

## Another alternative view: The Sanctity of Life view

There is another main standpoint here that we should consider before moving on. This is what I call the Sanctity of Life view. This view holds that *all* human lives are valuable to those who live them and thus would hold that we are never wrong to bring a life to birth, regardless of the expected quality of that life. This view might be based on a religious belief in the sanctity of all human life or simply a belief that human life is such a precious thing that it is assumed that all human life is something that should be valued. Those holding this view are likely to oppose the sanctioning of forms of assisted dying and the destruction of embryos and fetuses and might argue that we have a duty to reproduce.

This Sanctity of Life view is very different from the other views we have explored here. When faced with the question of whether human life is generally a good thing, those holding this view give their answer not based on the quality of life of individuals as the other views we have examined do. Instead, this view sees all or most human life as valuable with a right to life that comes from something outside the experience of individuals. This might be because human life is seen as a gift from God or similar.

## *Counterarguments to the Sanctity of Life view*

Arguing against what is often a religiously motivated position is difficult. Our usual approach of considering whether a position is based on accurate information or reasoning is difficult to apply when this is a position based on faith.

Those holding a Sanctity of Life view are likely to oppose procedures that cause the destruction of human embryos and fetuses. This is likely to mean that those holding these views would oppose the use of abortion and the use of embryos in experimentation or the development of treatments. This may also mean that those holding these views oppose the use of IVF and other fertility services that inevitably involve the destruction of embryos. It is also possible that those holding this kind of view might oppose routine screening for conditions like Down syndrome because of the close link between screening and termination of pregnancy.

If these views were used to shape regulation in a society, this would be problematic in terms of respect for individual autonomy, particularly the autonomy of individuals who can become pregnant. Inability to access safe abortion would lead to individuals having children they did not want to have or risking unsafe abortion procedures. If these views were used to prevent or significantly limit access to IVF, then the reproductive choices of individuals and couples would be affected. While holding these views as a private and personal choice is something that should be supported in terms of respecting individual autonomy in a pluralistic society, if we were to regulate on the basis of these kinds of views, this would be difficult to justify.

Further, this approach to the value of life is problematic for those of us who wish to take seriously the testimony of individuals. Listening to individuals' experiences of their own quality of life seems important when considering the ethics of reproduction. While the majority of people do experience their own lives as a positive experience, we know that there are some individuals who, for whatever reason, have lives that are not a positive experience. Giving higher consideration to something else but the self-reported welfare of individuals when taking a position on reproduction is one that is difficult to justify as a basis for regulation in this area.

## Where do you stand on the question: Do you think being alive is generally a good thing for most people?

Deciding where you stand on this question of whether you think that being alive is generally a good thing for most people is an important first step to evaluate the ethical questions we will explore in this book regarding how far we are justified in regulating individual reproductive choices.

I have argued that your intuitive response to this question is not a sufficient way to come to a robust position on this and other ethical questions. In order to answer this question in a way that is thorough and that you can feel confident about, I have suggested that you apply what I have called the 'ARC' approach, so that as well as providing your Answer to the question, you should also provide your Reason for your answer and deal with the main Counterargument to the position you take. As a result, when you consider whether you think being alive is generally a good thing, before you settle on your answer, it is important to ensure you have engaged with arguments on the other side of this debate so that you can be clear about your reasons for the position you take and you can defend this position against the main counterarguments. To give you an example of how this might work, my position on this question looks like this:

'Do you think being alive is generally a good thing for most people?'

*Answer:* Yes

*Reason:* When asked people invariably report that they value their lives, they feel glad that they are able to experience their lives and wish their lives to continue despite the negative experiences that are an unavoidable part of human life.

*What is the main Counterargument to my position?:*
It might be argued that while it seems that many or even most people value their lives all people suffer, and some suffer significantly even to the point that they might consider their own existence intrinsically harmful. As a result, it might be argued that being alive is not a good thing for anyone as it creates unnecessary suffering and risks creating lives of overwhelming suffering.

*My response to this counterargument to defend my position:*
While life can be intrinsically harmful for some this is not the case for most people. We know that most people will value their lives even if they face challenges in their lives. As long as we do our best to avoid bringing to birth children who will have intrinsically harmful lives, and we do our best to maximise the welfare of children who will exist, then those who want to have children should be supported to do so as being alive is generally a good thing for most people.

Of course, you do not have to agree with my position on this question. As you can see from my answer, there is often no perfect position on this and other ethical questions. While we might take a particular stance on this question, it is important that we take the other side of this argument seriously to ensure that our response can be defended. In this case, it is easy to see strengths in the arguments on both sides of this debate regardless of your final answer to the question.

Have a think about where you stand on this question and why and try and use the 'ARC' approach to develop your answer into one that you can explain and defend using the structure of the argument above to help you. Remember, it is absolutely fine to disagree with my answer and my reasons for my answer.

## Your answer to this question and its influence on the three questions that are the focus of this book

The way we answer the question 'Do you think being alive is generally a good thing for most people?' will have a strong influence on how we answer the three main questions that are the focus of this book:

- *Question 1: Are we justified in attempting to evaluate the potential parenting ability of those trying to access fertility treatment (e.g., disabled people or individuals with past criminal convictions) and prevent access in some cases?*

- *Question 2: Should we allow prospective parents using IVF to implant an embryo with a condition considered a disability? For example, should a deaf person be allowed to implant a 'deaf' embryo?*

- *Question 3: Is routine screening for Down syndrome in pregnancy ethically acceptable even if there is evidence that individuals may feel pressure to accept this screening?*

There will be many different versions of the general views we have discussed above. There will be those of you who arguing that we should be discouraging reproduction generally and others who believe that having children is always a good thing, no matter their expected quality of life. While I do not simply dismiss these views, these perhaps more polarized views would not be expected to argue for the sort of regulation in reproduction based on the welfare of the child that we focus on in this book. This is because at one end of these divergent views, the Antinatalist view, the welfare of future children always represents an unacceptable risk. At the other, the Sanctity of Life view, the welfare of future children is invariably judged to be positive.

Others holding an Antinatalist view might, however, argue for a more moderate view of Antinatalism. On this more moderate view, it might be argued that overriding the reproductive choices of those who need help with reproduction is unjust, even if reproduction *per se* is difficult to defend. They might be motivated to try and persuade others of their views rather than to interfere with reproductive choices directly. Regulation based on this more moderate form of Antinatalism might look more like my point of view than we might first expect. If someone with this moderate view of Antinatalism thought that assisted reproduction should be available on the grounds of equality, they might well wish to still include guidance and safeguards about protecting the welfare of future children given that most of the arguments that motivate the Antinatlist view are focused on concern for the welfare of future children.

Similarly, a more moderate Sanctity of Life view might be one that would tolerate assisted reproduction despite the issues around the destruction of embryos in order to allow more individuals and couples to have children, something that, on this view, is a really good thing. If this view were taken, it might also be that this moderate Sanctity of Life view might involve some consideration of the welfare of future lives created as part of this process.

Thus, unless you take the view that either reproduction is always a bad thing or that reproduction is always a good thing, you will probably take a view that is likely still concerned with developing regulations protecting the welfare of future children.

As we have seen, one way of thinking about this is by trying to determine which lives are likely to be intrinsically valuable to those who experience them and which lives fall below this threshold and are likely to be lives that are intrinsically harmful to those who experience them. Based on this approach, if we estimate that we will be creating a life that is likely to be an overall harm to the person living it, something overwhelmingly bad for that person, then it seems that, in these circumstances, we do have a reason to attempt to influence the reproductive choices of individuals which are likely to cause those with such harmful lives to exist. As we have seen, however, if we take the view that being alive is generally a good thing, even though this inevitably entails significantly negative aspects, then unless a life is dominated by negative experiences, it is difficult to make a case that this life will be experienced as a harm by the person who lives it.

Based on this argument, we might have a good reason to enact regulations that do not permit individuals to implant IVF embryos with conditions that we judge are likely to render a life intrinsically harmful to those who experience it. Similarly, we might have good reasons to avoid providing fertility treatment to individuals where we estimate that the resulting child will have a life of overwhelming suffering and little of positive value. We might also be justified in developing screening programmes to identify and provide information to pregnant people about catastrophic conditions that are likely to result in lives that we expect to be intrinsically harmful for those living them.

But, of course, this approach relies on us being able to draw a line between what we consider to be intrinsically valuable and intrinsically harmful lives. So the next job we need to do is to start examining this distinction further.

## Are the lives of the future children in the cases we are considering likely to be intrinsically valuable or intrinsically harmful lives?

If we accept that there are some lives that we might consider intrinsically harmful and thus that we have a moral obligation to attempt to avoid creating, we then need to consider the question:

Is the particular choice we are considering likely to produce an intrinsically harmful life?

When considering cases where a child has not been conceived or born yet, we need to consider whether the seemingly suboptimal conditions of this particular child's birth are such that they will experience this life as intrinsically harmful. When we consider cases where a child will be born with deafness or Down syndrome or perhaps to a parent who has a disability, we need to consider whether these kinds of conditions are such that they are likely to render the resulting person's life a harm overall or whether those with deafness, Down syndrome or those with parents that might find parenting more challenging than others[32] are as likely to value their own lives as any others might. We all face challenges in our lives: our looks or our height might make our lives go worse than they might have, our gender, race or religion might make our lives more challenging than they could have been, or our social background or LGBTQI+ identity might also mean that we face obstacles and difficulties that others do not. None of us have perfect parents, and most of us have parents who make mistakes or face their own challenges when it comes to bringing up children. Many parents work too much, do not spend enough quality time with their children, put pressure on their children to succeed and invariably will at some time lose their temper or act in ways that undermine a child's self-esteem. While none of these attributes of parents are ideal, they are unlikely to have such an impact on a child that they no longer value their life and wish it to continue. So while bringing to birth children where we can predict that there will be challenges to their lives can feel uncomfortable in many cases, we need to examine this discomfort and consider it against the background that all lives have challenges and that if we hold that being alive is generally a good thing, then creating challenging lives that we nevertheless expect to be lives that those living them value, then it is very difficult to see why doing so is morally questionable.

## Different people choices

To understand why it is difficult to show reasons, based on the welfare of the future child, to justify the sort of regulations that we are considering in this book, we first need to recognize that these are choices about *who* should be born and not choices about in what conditions we should allow someone to be born in. These are different people choices rather than choices about different circumstances for the same individual.

We are used to assessing the welfare of existing individuals in terms of comparisons. When considering choices regarding existing individuals, we think about whether our decision will make them better or worse off than they otherwise would have been. For this reason, the question that many of us ask in these circumstances is often, 'Would this child be better being born without a disability or perhaps to parents without the challenges faced by these parents?'. While this would be the appropriate question when considering decisions that would affect an existing child, the fact that we are making decisions about who will be born and considering lives not yet started makes these kinds of comparisons not applicable.

We will explore this comparative notion of harm and benefit and how this may affect our thinking about these questions in more detail later. However, for now, it is important to recognize that the decision being made here will bring to birth this particular child, no child at all or a different child. This particular child can only be born in the condition that they can be born in and to these particular parents; thus, this comparative question is not applicable in these cases. The choice that is available to us is whether this child will come to exist or not. This fact is fundamental to thinking through the decision we make here. We are not choosing whether a particular child will have what we consider to be extra challenges or not, but we are considering whether we do something wrong to allow or enable this child, who can only be born with these challenges, to be born. Our answer to this question will depend upon whether we consider the conditions of this child's existence to be likely to be such that they render that life a harm in itself to the child in question. Thus, the pertinent question here is, 'Do we do something bad or harmful to this particular person by allowing them to exist with this particular condition or in these particular circumstances?'.

## Conclusion

Deciding where we stand on this question of whether we think life is generally a good thing for most people is an important starting point when addressing the sort of questions that are the focus of this book. Thus, spending a bit of time to not only decide what your intuitive answer is here, but also *why* you hold this answer and considering the contrasting responses to this question and why, for you, they are not convincing, is time well spent. There will be many of you who, after this process, take or continue to take a view, like the Antinatal view or the Sanctity of Life view, which might mean that the questions we focus on here are less immediate questions for you when it comes to the ethics of reproduction.

However, for those of us who think that being alive is generally a good thing for most people, we are, I suggest, left with some interesting assumptions but ones that lead to what many find to be an intuitively uncomfortable conclusion:

a) The majority of people have what we consider to be intrinsically valuable lives. These are lives that are such that the balance of positive and negative experiences make them something overall that those living these lives value, do not regret and wish to continue living.

b) While being alive is usually (on balance) a good thing, there are lives that are intrinsically harmful to those who experience them. These are lives where negative experiences overwhelm any positive experiences and make these lives ones that we might expect those who are living them not to value, to regret and to wish to end.

c) If a person is brought to birth with an intrinsically valuable life, that is, a life that he is likely to value, not regret and wish to continue, then bringing that person to birth is not a harm for them.

d) It is wrong to deliberately harm a child by making it worse off than it might have been, but this is not the situation with these choices. These are different person choices where the choices we are making is whether it is acceptable to allow this child to be born in the only conditions or into the only conditions they can be born in and with.

e) Unless these conditions are so bad as to overwhelm a life with negative experiences, then this individual is as likely as any of us to have a life that they would consider to be intrinsically valuable, and we cannot conclude that this is bad for the child and thus that this choice is a morally unacceptable choice. As a result, we cannot justify attempting to influence or prevent choices that would bring intrinsically valuable lives to birth, even if these lives are ones that we might consider to be 'suboptimal' or disabled.

f) While we might have good reason to attempt to influence or prevent the birth of those whom we might consider will have intrinsically harmful lives, lives that fall into this category are likely to be rare cases where negative experiences overwhelm any positive experiences.

g) As a result, we can conclude that in most cases, we do not have reasons, based on the welfare of the resulting child, to attempt to prevent the birth of this child.

Many people, while accepting the above assumptions, find the conclusion that we have no good reason to oppose choices to have children who, we may feel, face considerable challenges, difficult to accept. In the next chapter, we examine these assumptions and conclusions further and explore why we might feel uneasy about these conclusions.

# CHAPTER 4
# WHY DO I FEEL UNCOMFORTABLE ABOUT THIS CONCLUSION? THE NON-IDENTITY PROBLEM

In the previous chapter, we considered the different approaches to the question of when bringing a child to birth might be harmful. Unless we take the view that all reproduction should be avoided where possible or that all reproduction is inherently a good thing, we seem to be left with the conclusion that an individual who is born deaf or with Down syndrome or born from IVF with what might be considered suboptimal conditions (perhaps with a parent who has their own health challenges) is born in the only condition they can be born in and is as likely as anyone else to value their own life. Given that these individuals are as likely as any of us to value their own life, this seems to lead to the conclusion that we do not do anything wrong to this person by causing them to exist. Consequently, it is difficult to see how we can justify interference in reproductive choices based on harm prevention, as most reproductive choices do not seem to harm any individual.

However, while this argument, when set out in this way, may seem to have a lot of merit, one of the most interesting philosophical debates in this area surrounds the fact that even if we accept the positions that make up this argument, many of us still feel uncomfortable with the conclusion these positions seem to lead to.

## The Non-Identity Problem

This feeling of unease about the rightness of choosing to have a child who is likely to have more challenges than an alternative child, even if it is difficult to identify any harm caused by this choice, is known as the Non-Identity Problem and was first put forward by Derek Parfit.[1]

To understand the Non-Identity Problem, let's look at our Question 2:

*Should we allow prospective parents using IVF to implant an embryo with a condition considered a disability? For example, should a deaf person be allowed to implant a 'deaf' embryo?*

Imagine that prospective parents using IVF are able to identify a 'deaf' embryo using pre-implantation genetic testing and choose to implant an embryo that will become a deaf person rather than choosing one that will be a hearing person. Assuming we do not take an Antinatlist view or a Sanctity of Life view of reproduction, in response to this scenario, we might adhere to the following statements:

- This choice changes the identity of who will be born. A deaf child will be born instead of a hearing child. So here we are talking about two different possible individuals.

- No one is harmed or made worse off than they could have been by this choice – the deaf child can only be born as deaf and not as hearing.

- If the deaf child is brought to birth, they are as likely as anyone else to have a life they will value.

- It is difficult to conclude that this choice to choose the 'deaf' embryo is a morally unacceptable choice as it is not 'bad for' the child created.

- However, many of us may still have a strong feeling that making this choice is morally unacceptable.

This is the Non-Identity Problem. We appear to have shown that no one is harmed by choosing to bring life with a disability or other foreseeable challenges so long as the life created is expected to be intrinsically valuable. As no one is harmed, it seems reasonable to assume that this is not a bad thing to do, but we feel uncomfortable with this conclusion.

It is this discomfort with the conclusion here that makes this the Non-Identity *Problem* and not the Non-Identity *Argument*. Unless you take an Antinatalist view, very few people would argue that those with disabilities such as deafness or Down syndrome and those born to parents with disabilities or other challenges are likely to have a life that is intrinsically harmful to those who experience these lives. Empirical evidence shows that when we ask people with these kinds of lives, they are as likely as anyone else to value their lives.[2] We recognize that these kinds of lives can have the same spectrum of quality as any other lives.

While it is difficult to argue against the testimony of individuals with these conditions or challenges about the quality of their own lives, many still feel that choosing to bring someone to birth with these disabilities or challenges is ethically unjustified and even morally wrong. Savulescu and Kahane, for instance, argue that 'it is in fact implicit in commonsense morality that it is morally permissible and often expected of parents to take the means to select future children with greater potential for well-being'.[3]

This sentiment echoes the sentiment of many others to whom it seems obvious and unquestionable that preventing someone from choosing to have a disabled child or even being selective about who should get fertility treatment when there are concerns about the welfare of the future child are the 'right' things to do if we care about the welfare of children. The phrase 'commonsense morality' seems to express this idea that it just seems obvious that to choose to bring to birth someone who is disabled when a non-disabled child could be born instead is wrong.

Parfit shared this intuition that choosing to create what he considered to be lives with greater challenges rather than lives without these challenges felt strongly intuitively wrong. As a result, he presented the Non-Identity Problem as a philosophical puzzle that he felt needed to be solved in order to account for the unease we feel about what he called the 'disturbing conclusion'[4] that if no one is harmed by the choice to bring to birth a life with foreseeable challenges, then it seems that this is not a morally wrong thing to do. He argued that what was needed was a 'Theory X'[5] that would show what was wrong with the reasoning of the Non-Identity Problem and thus allow us to retain our intuition that choosing to bring to birth a more challenged life than the alternative is a morally questionable or even morally wrong choice. While many people have tried to find the 'Theory X' that identifies a flaw in this reasoning, attempts to do so must make sense of why a choice that appears to harm no one is wrong, which is extremely difficult.

If we cannot find a problem with the reasoning here, then we seem bound to accept the conclusion that choosing to bring a disabled child or a child to birth in what may be seen as challenging circumstances harms no one and thus is not wrong. If we accept this conclusion, we appear to have no justification for regulations that make choosing to have a disabled child who is likely to have an intrinsically valuable life difficult for those who wish to make this choice.

I argue that we should accept the conclusion of the Non-Identity Problem, that choosing to bring disabled but intrinsically valuable lives to birth does no harm. As a result, I further argue that these choices cannot and should not be condemned, discouraged or prohibited. I suggest that our discomfort about this conclusion can be explained in various ways, including the general bias around disability and our assessments of the quality of disabled lives. More of this later, but first, we will consider some ways people have tried to reject the conclusion of the Non-identity Problem and maintain the idea that choosing to bring to birth a disabled life is morally wrong.

## An obligation to bring to birth the best child possible?

If we want to argue, as many do, that we have a moral obligation to try and bring to birth the best child possible and to avoid creating disabled children where possible, then we have to do more than rely on intuition. Claiming that avoiding bringing children with disabilities to birth is morally wrong is a claim with serious consequences. This claim implies that those living with disabilities are less valued.[6] It has the potential, particularly if unjustified, to perpetuate negative attitudes towards these conditions and those who already live with them. Further, even though those who propose the existence of an obligation to bring to birth the 'best' child possible often argue that this obligation should not be enforced,[7] claims by highly respected academics that chime with widespread intuitions are extremely likely to give weight to regulations that may be used to enforce this supposed obligation.

Reinforcing negative attitudes towards conditions viewed as disabilities and seeming to provide academic justification for an intuition that can be used as the basis for regulation that aims to reduce the incidence of those born with these conditions might also be argued to be eugenics.[8] We will explore this claim further later. But for now, I argue that unless these intuitions can be justified by an argument that allows us to reject the conclusion of the Non-Identity Problem convincingly, then we are in danger of basing regulation with severe consequences on no more than intuition. As we have seen, history teaches us that basing regulation on intuition or public feeling alone is a dangerous approach if we care about the equal rights of individuals.

There have been a number of people who have tried to provide a justification for this idea of a moral obligation to bring to birth the 'best' child possible. Parfit was one of the first to discuss such a possible obligation.[9] Harris brought these arguments into the bioethical and applied ethics arena around the same time,[10] arguing that 'it may be morally wrong to "choose" to bring to birth an individual with any impairment, however slight, if a healthy individual could be brought to birth instead'.[11] Later, Savulescu suggested the label 'the Principle of Procreative Beneficence' for this claim[12] and argued that where choice is possible, for instance, when selecting IVF embryos for implantation, the Principle of Procreative Beneficence requires that

> couples (or single reproducers) should select the child, of the possible children they could have, who is expected to have the best life, or at least as good a life as the others, based on the relevant, available information.[13]

However, to justify this claim that there is a moral obligation to bring to birth the 'best' child possible, those supporting this claim need to show why the conclusion to the Non-Identity Problem does not hold.

A huge amount of work has gone into attempting to solve the Non-Identity Problem and thus justify the existence of a moral obligation to avoid creating disabled lives where possible. It is impossible to explore the details of these arguments in the space I have within this book. However, these potential solutions roughly fit within three different categories:

- We accept the conclusion of the Non-Identity Problem and reject the notion that we have a moral obligation to bring to birth the 'best' child possible. Taking this strategy will involve trying to explain why we often feel so uncomfortable about the conclusion to the Non-Identity Problem.

- The second strategy is to reject the idea that something wrong or harmful must be bad for a particular person. This strategy invokes the notion of non-person-affecting or impersonal harm.

- The third main strategy is to find a way to reject the idea that if a person is brought to birth with an intrinsically valuable life, then bringing that person to birth is not a bad thing to do. The argument here is that we can reject the conclusion of the Non-Identity Problem because the choice to bring to birth a disabled life, even if it will be intrinsically valuable to the person who lives it, is a choice that harms or wrongs that person.

I take the first of these strategies, and I will explore this further in the next chapter. But for now, we will explore the other two main strategies of trying to solve the Non-Identity Problem, starting with attempts to solve this puzzle by appealing to the concept of impersonal harm.

## *Solving the Non-Identify Problem by an appeal to impersonal harm?*

There are a number of reasons why establishing an obligation to bring to birth the best child is problematic. For instance, just deciding on what we mean by the 'best' child is highly subjective, and thus, applying this principle with any consistency would be practically challenging.[14] However, despite this and other issues with these arguments, the main challenge to providing a convincing argument in favour of this moral obligation is explaining why choices to bring a disabled child to birth are morally wrong when this choice harms no person.

The main way in which Parfit, Harris, Savulescu and others attempt to justify the existence of this moral obligation is by providing examples of scenarios. For instance, Parfit introduces the Non-Identity Problem by asking us to consider his 'Risky Policy' example. The Risky Policy example asks us to suppose that we must choose between two energy policies. Both policies would be safe for at least three centuries. However, one policy, the Risky Policy, as well as increasing the quality of life for the population, would also entail a small risk of radioactive contamination in the future. The Risky Policy is chosen, and many hundreds of years later, radiation is leaked, causing the premature deaths of thousands of people. We instinctively feel that choosing the Risky Policy is wrong. But Parfit argues that the wrongness of choosing the Risky Policy cannot be explained in terms of person-affecting consequences, that is, harm to particular individuals.

Parfit explains that given the infinite number of variables that have to be in place for a particular egg and sperm to fuse and create a specific individual, any changes in society are likely to change who will be born. As a result, he argues, the individuals who suffer premature death because of the adoption of the Risky Policy cannot be said to have been harmed by this Policy as it was the particular conditions that happened to be in place, including the increased quality of life brought about by this policy, that meant that they were conceived rather than someone else. While most of us would see the choice of the Risky Policy as the morally wrong choice, given that those affected were born in the only condition they could be, and given that they would be expected to have lives that they value, is it not clear that we have harmed these individuals, even though their lives are shortened.

Parfit uses other examples focusing on reproduction rather than policy decisions. For instance, he asks us to consider the actions of a fourteen-year-old girl who chooses to conceive a child, knowing, Parfit argues, that 'this will have bad effects throughout this child's life'.[15] These effects are not such that they are likely to render their life intrinsically harmful but will, he argues, cause this life to go worse than another alternative life could go if she chose to wait.[16] This and other reproduction examples are close to the ones that we have been considering. If we hold that being alive is a generally good thing for most people, then these reproductive cases seem to indicate that, as with the Risky Policy, we harm no one by making this choice, but many of us will still feel uneasy about this choice.

This device of using scenarios like those above to initiate the intuitive negative response that many of us have to these choices is the main way in which those arguing for an obligation to choose against disability and for the 'best' child possible back up their claim. Harris, Savulescu and others use variations on these scenarios with which they hope to give weight to their claim for this moral obligation to avoid creating children with disabling conditions. However, without invoking something else here, these scenarios seem only to confirm the Non-Identity Problem, that is, that while we do not harm anyone by this kind of choice, this choice nevertheless produces a strong intuitive feeling of unease in many of us.

Parfit explores the idea that the 'something else' we could be looking for here, the Theory X, that allows us to reject the conclusion of the Non-Identity Problem, is something he calls impersonal harm. The notion of impersonal harm rests on the idea that having a life of overall positive value is a good thing, and the better the quality of life one has, the higher the value this life has. Parfit calls this the *Impersonal Total Principle*, saying, 'If other things are equal, the best outcome is the one in which there would be the greatest quantity of whatever makes life worth living'.[17] Those who put forward this idea of impersonal harm usually do so on the basis of utilitarianism, an approach to ethical problems that suggests that we should attempt to maximize happiness, well-being or whatever we think makes human lives go well. Based on this idea of maximizing happiness or well-being, the argument here is that the better the quality of life of the individuals who live, the bigger the total amount of this happiness or well-being in the world. Thus, the impersonal harm caused by choosing to bring to birth a child with a disabling condition or in challenging circumstances is that it reduces the total amount of happiness or well-being in the world.

While, as we will see in a moment, Parfit is unable to provide a conception of this impersonal harm that stands up to scrutiny, others who champion this idea of an obligation to bring to birth the 'best' child possible do so based primarily on this difficult concept of impersonal harm. For instance, Harris argues that while we haven't done anything wrong[18] to the person who is born with an impaired life, as they are likely to have a life they value, a choice to bring to birth a disabled or disadvantaged life is still wrong because it makes the 'world a worse place than it need have been'.[19] Similarly, Savulescu argues that in order to justify his arguments that it is 'bad that blind and deaf children are born when sighted and hearing children could have been born in their place',[20] we must 'appeal to some form of harmless wrong-doing, we must claim that wrong was done, but no-one was harmed'.[21] Both Harris and Savulescu are appealing to a non-person-affecting harm or what Parfit calls impersonal harm to attempt to justify their arguments.

### *Problems with the notion of impersonal harm*

Parfit was one of the first people to explore this concept of impersonal harm and how it might be used to solve the Non-Identity Problem in the 1970s.[22] However, he was unable to come up with a conception of impersonal harm that stood up as a plausible concept. As you will remember, arguments based on the concept of impersonal harm rely on there being a motivation to increase the cumulative totals of happiness or whatever makes life good, and this motivation, if accepted, leads to some rather unpalatable conclusions.

Firstly, taking seriously the importance of impersonal harm seems to imply a moral obligation to create as many intrinsically valuable lives as possible as doing so will increase these cumulative totals of happiness or well-being very effectively by simply increasing the number of people who exist and who are likely to value their lives. However, the notion of having a moral obligation to reproduce and have as many children as possible is hugely counter-intuitive and would seem to have serious implications for the autonomy of, in particular, those who would be required to bear these children.

Secondly, it seems that if we were to take this obligation to maximize the creation of intrinsically valuable lives seriously, increasing the population is likely to eventually decrease the average quality of life dramatically due to overcrowding and scarcity of resources. As Parfit puts it, 'For any possible population of at least ten billion people, all with a very high quality of life, there must be some much larger imaginable population whose existence, if other things are equal, would be better [in terms of cumulative totals of happiness/well-being], even though its members have lives that are barely worth living'.[23] If what motivates us is increasing these cumulative totals of happiness or well-being, then this could be achieved by massively increasing a population. This massive increase in population is likely to dramatically decrease the quality of life of those in this population. But as long as those in this larger population have intrinsically valuable lives, this will lead to a larger cumulative total of happiness or well-being. Thus, a choice to increase cumulative totals of happiness or well-being by increasing the population won't improve the welfare of anyone and may well lead to a lower quality of life overall. This is what Parfit calls the Repugnant Conclusion. What makes this conclusion repugnant is that choices to increase the population in this way seem to ignore what we usually see as important: the welfare of individual people.

If we accept the idea of impersonal harm as something that should influence our choices, we are then committed to highly counter-intuitive motivations that focus not on the welfare of individual people, but on maximizing the cumulative totals of happiness or well-being. If we were to make decisions based on this maximization, then it is unlikely that we would increase the welfare of any particular individuals, and there is a real danger we would decrease the welfare of some people. As such, this notion of impersonal harm seems to be a very counter-intuitive way of dealing with an intuition many of us have about why choosing to bring to birth an intrinsically valuable but disabled or disadvantaged life might be a bad thing to do.

Further, and perhaps most problematically, deciding *what* to measure here, when it comes to these cumulative totals of well-being, is highly subjective and open to bias, leaving us to ask the question, 'What does the 'best' life even look like?'.[24] What kind of life gives the highest score on this idea of cumulative totals of what makes life go well? Is physical perfection the most important factor, or intelligence or longevity? If something like well-being or happiness is the element to be measured, how do we calculate this?

If we base these cumulative totals of what makes life go well on the testimony of those who live those lives, we may well find that it is not what we consider to be disabilities or disadvantages that necessarily make lives have a lower quality overall than others. It may well be that social conditions have an equal or greater impact on the welfare of individuals.

If we base these cumulative totals of what makes life go well on a lack of what are considered to be physical or mental disabilities, then prioritizing these cumulative totals of welfare might lead us not only to attempt to screen out all 'disabling' conditions but other conditions or characteristics that might be deemed to make lives more challenging than others: for instance, individuals with LGBTQI+ identities, those with autism, those considered by cultural norms to be less attractive or others with characteristics that may add challenges to lives.

## *Eugenics*

It is not surprising that those who put forward this idea of a moral obligation to bring to birth the 'best' child possible have been accused of promoting eugenics.[25] The claim made is that we have a moral obligation to bring to birth the 'best' children possible, to avoid bringing to birth children with disabilities and 'disadvantages' where possible. This claim is justified not by the welfare of individuals but by a desire to make a better world in some way that is difficult to define.

Savulescu has attempted to distance the Principle of Procreative Beneficence (the term he uses for this obligation to bring to birth the best child possible) from the notion of eugenics, saying,

> [...]Procreative Beneficence is different to eugenics. Eugenics is selective breeding to produce a better *population*. A *public interest* justification for interfering in reproduction is different from Procreative Beneficence which aims at producing the best child, of the possible children, couple could have. That is an essentially private enterprise. It was the eugenics movement itself which sought to influence reproduction, through involuntary sterilisation, to promote social goods.[26]

It is difficult, however, to see how such a defence can be maintained. It seems that the Principle of Procreative Beneficence and the moral obligation it represents is the very embodiment of Savulescu's definition of eugenics. Based on this definition, eugenics is a '*public interest* justification for interfering with reproduction' 'to promote social goods'.[27] As we have seen, the establishment of a moral obligation to bring to birth the best child we can is not built on the private interests of the prospective parents regarding what sort of child they wish to have, as fulfilling this obligation will restrict some reproductive choices. This obligation is also not founded on the individual interests of the child who will be created, as their welfare will not be affected by the decision about which embryo to implant or which pregnancy to continue. What this obligation is built on is an idea of making the world a better place than it could otherwise have been, in terms of creating the greatest cumulative totals of whatever it is decided makes life go well.

If a project is not interested in the welfare of particular people but in creating what those proposing this project believe is the best world possible, then this is exactly what eugenics is – promoting social and not personal goods. It is true that those proposing this obligation to create the best children possible do not advocate state sanctioned coercion at the level of forced sterilizations or terminations to achieve this end. Still, it is not true that no coercion at all is implied. Backing from high-status academic 'experts' is likely to provide significant support for those who wish to enact or maintain regulations that rest on this idea of this obligation to bring to birth the 'best' child possible. These regulations, which seem to confirm the social norms and intuitions around this moral obligation to choose to bring to birth the 'best' child possible, will inevitably result in pressure to meet this supposed obligation.

Harris does not shy away from the charge of eugenics as Savulescu does. Harris states, 'I specifically adopt the Oxford English Dictionary definition of eugenics as "pertaining … to the production of fine offspring" and say that if this is what eugenics is everyone should favour eugenics'.[28] If he is right about his argument that 'deliberately to make a reproductive choice knowing that the resulting child will be significantly disabled is morally problematic, and often morally wrong',[29] then it would seem that publicizing this moral obligation and allowing it to influence regulation in order to enable its fulfilment would seem to be acceptable, even if such policy could then technically be termed eugenic. However, if we cannot provide reasons for this moral imperative and fall back on intuition and cultural norms for justification, then regulation based on this supposed moral obligation is eugenics at its most objectionable.

### *The notion that impersonal harm inevitably places a lower value on the disabled*

There is a further problem with the notion of impersonal harm. This is that it inevitably places lower value on disabled people, which is not only unwarranted and offensive, but is also likely to reinforce biases and discrimination against those living with conditions considered to be disabilities. If we think about the utilitarian calculation that has to be made to try and explain why a world without disabled people is a better world than a world with disabled people, we can only reach this conclusion if we accept that the lives of disabled people have less value in this calculation It is only if these lives count for less in this calculation of cumulative totals of well-being that those using these arguments can show that the world where choices are made to bring to birth disabled people is worth less and is, therefore, a worse world than the alternative.

Those who propose this moral obligation based on the notion of impersonal harm would object to this conclusion that this argument necessarily places lower value on disabled people. Harris, for instance, stresses that 'all persons are equal and none are less equal than others. No disability, however slight, nor however severe, implies lesser moral, political or ethical status, worth or value'.[30] However, it is the assumption that this world with less disabled people, where individuals choose against disability, is a morally better world that is problematic here.

If it were the case that disabled people were significantly more likely than others to experience life as a harm, that is, as something that is overwhelmed with suffering or other negative factors, then we might have a reason to say that these lives are morally undesirable and choices to create these lives should be discouraged. However, this is not the case. As a result, the only way we can understand that the lives of disabled people count for less when it comes to ideas of cumulative total well-being, or impersonal goods, is if we reject the assessment of the quality of these lives that those people living them provide and replace this with the judgement of others who have a strong belief that while disabling conditions are usually compatible with inherently valuable lives, there are still reasons why creating these lives is not just undesirable but also immoral.

As we will see later, I argue that our intuition regarding this obligation to choose to bring the 'best' child possible to birth is motivated by our preferences around disability that are hugely influenced by bias and cultural norms. As we have seen, the suggestion that this intuition is more than this and indicates a moral obligation cannot be justified by appeals to the difficult concept of impersonal harm. Thus, unless we can find another concept that gives us a convincing way of rejecting the conclusion of the Non-Identity Problem, we are left with an argument that cannot be justified or defended and that has serious consequences in terms of infringing on the reproductive choices of individuals and reinforcing bias and discrimination towards those living with conditions that may be viewed as disabilities.

### *Limited Impact*

Even if we were somehow able to overcome these many challenges to this notion of impersonal harm to justify an obligation to avoid what are seen as suboptimal lives, this would still not provide a strong justification for the regulations we have been considering in this book. Even if it could be established that we have a moral obligation to ensure we do not make this 'world a worse place than it needed to be'[31] it is very unlikely that removing regulations that influence reproductive choices would significantly affect cumulative totals of welfare. This is because it is very likely that allowing this reproductive freedom will only affect a small number of prospective parents and, thus, allowing these choices will be very unlikely to make a significant change in the population and any cumulative calculations of well-being. As a result, even without the theoretical challenges that this concept presents and the problematic and very subjective value judgements it makes about different kinds of lives, we are still on very shaky ground attempting to argue that a world with reproductive freedom to create intrinsically valuable lives is a 'worse' world than an alternative world.

## Other possible solutions to the Non-Identity Problem: The disabled life is one that harms or wrongs the person who experiences it

As we saw earlier, while this area of debate is complex, there are three main strategies when it comes to attempting to solve the Non-Identity Problem. We can accept the conclusion of the Non-Identity Problem and provide an explanation for our feelings of unease about the conclusions. We can reject the idea that for something to be bad, it has to be bad for someone and posit the existence of non-person-affecting harm. Our third strategy is to try to retain the person-affecting idea of the badness of the choice to bring to birth a disabled but intrinsically valuable life by arguing that such choices harm or wrong the resulting person. Here, the suggestion is that those born with disabling conditions are harmed or wronged by being brought to birth, and thus, these choices should be avoided where possible. Below, I will outline some of the possible arguments here, but given the space constrictions, this can only be an overview of some of these arguments.

### Antinatalism and harm

The first of these arguments is one we have already come across. This argument comes from those who take an Antinatalist stance. It is argued that all human life is harmful; thus, we are all wronged by the decision to bring us into existence. The argument here is that those of us who insist that we value our lives are simply mistaken about this and have an unrealistic perception of the suffering that is integral to human life.[32] It seems that while this might be an argument for the immorality of reproduction generally, it cannot be an argument that solves the Non-Identity Problem. If we accept that all lives are harmful, then this gives us reason to avoid the birth of everyone. It also seems difficult to square this argument with the testimony of most human beings who, when asked, are clear that they experience life as a good thing.

A slightly modified version of the Antinatalist argument here might be that given that life is generally bad for those who experience it, we should try and ensure that if we are to create this avoidable suffering, we should try and create as little suffering as possible by trying to create as good a life as we can and avoiding obvious disabilities and 'disadvantages' that are thought to make a life go worse.[33] Again, it is difficult to reconcile this argument with the testimony not only of human beings generally, but also of those who live with the sort of disabilities and 'disadvantages' that we are discussing here. It seems that those living with deafness or conditions like Down syndrome or with parents who might have more challenges than others are as likely as anyone else to value their life. Even where this is not the case, it may be that the things that we see as disabilities or 'disadvantages' are not the things that those living these lives find challenging. It may be, for instance, that for some deaf individuals, it is the attitudes of the hearing community or the lack of reasonable adaptations that increase the challenges in their lives.[34] Or it might be that poverty or other factors are a greater source of negative experience than the conditions others may see as disabilities or 'disadvantages'.

## Harris and harm

Another example of an argument that rejects the idea that those born with disabled but intrinsically valuable lives are not harmed comes from Harris. As well as arguing for impersonal harm, Harris argues that bringing to birth a disabled child is wrongful as it 'causes a child to be born in a 'harmed' condition'.[35] Harris accepts that the disabled child has 'no complaint because for them the alternative was non-existence'.[36] Here, he argues that while we do nothing wrong to this child as they are likely to have an intrinsically valuable life, we have unnecessarily harmed this child.

Harris argues that we have 'a strong moral obligation to prevent preventable harm and suffering and that this obligation applies equally to curing disease and injury and to preventing the avoidable creation of people who will have disease or injury'.[37] Harris argues that those born with a disability have been harmed and suffer from this harm. For instance, Harris writes, 'I do not believe there is a difference between choosing a pre-implantation deaf embryo and refusing a cure to a newborn. Nor do I see an important difference between refusing a cure and deliberately deafening a child'.[38] Harris' argument here is that if we think that deafening a hearing child or denying them a cure for deafness is wrong, we do so because being deaf is harmful, and people suffer from their deafness. In turn, he argues that those born congenitally deaf also suffer from this lack of hearing and are harmed by this choice to bring them to birth.

However, I argue that there are serious problems with Harris' reasoning here. These cases are not comparable. While deafening a hearing child may harm that child, this harm makes sense as this action may make this individual worse off than they could have been. However, a choice to bring to birth a deaf child does not make anyone worse off than they could have been; this child could only be born deaf or not at all.

As we will see in the next chapter, when we examine this idea of comparative notions of harm and benefit, Harris' protective instincts may be kicking in here when he is concerned about the birth of a child with congenital deafness. But in this case, there is nothing that we can or need to protect this child from. They are either born deaf or not at all and being born deaf is unlikely to cause their life to be overwhelmed by negative experiences. Given that this congenitally deaf child is not made worse off than they could be by the choice of bringing them to birth and Harris agrees that they have 'no complaint' as they have a life that they are likely to value, it is very difficult to make sense of the harm that is being argued for here.

Harris' argument has a further problem shared with our previous Antinatal arguments, that if what motivates us is to avoid avoidable harm, and Harris is arguing that even those with intrinsically valuable lives are harmed by the choice to bring them to birth, then this seems to render all reproduction ethically unacceptable. All lives cause the person to suffer through pain, disease, injury, frustration and loss. If we are to deny that having an overall intrinsically valuable life compensates for this suffering, then it seems we are all harmed in the way that Harris argues that those born with disabilities like deafness are harmed.

### *Personal identity and harm*

One of the assumptions leading to the Non-Identity Problem is that the choices we are considering in cases affected by the Non-Identity Problem are different person choices. For instance, in the case of deafness, if we wanted to choose to bring to birth a deaf child rather than a hearing child, we are not suggesting deafening a child. We are talking about conceiving a different child, perhaps by using a congenitally deaf sperm donor or implanting a different 'deaf' embryo to the alternative hearing ones.

The result of this different person choice is that, so long as those chosen are likely to have an intrinsically valuable life, it is difficult to understand who has been harmed by this choice. Whichever way we choose, someone is expected to live a life of intrinsic value, and no one is made worse off than they could have otherwise been. This is based on what Parfit calls the *Time Dependence Claim*. This is the idea that '[i]f any particular person had not been conceived when he was in fact conceived, it is *in fact* true that he would never have existed'.[39]

We assume that our personal identity is shaped significantly by our genetic identity. That is, in order for a particular person with a particular identity to come into existence, a particular sperm and egg would need to come together to create that specific genetic individual. This view is often referred to as 'material-origin essentialism'[40] as it is the idea that one's genetic origins are essential to the identity of an existing individual. On this view, an individual born with a congenital disability or with other characteristics, including their sex and race, could not be born with other characteristics. However, some anti-essentialists[41] argue that by focusing on genetic identity, we are taking too narrow a view of personal identity. The claim here is that if we take a wider view of personal identity, we can show that this person is made worse off by choosing to have a deaf rather than a hearing child.

The suggestion here is that the genetic view of personal identity is not the only way of looking at personal identity. We know that there are many other factors that affect our personal identity, including the environment we are brought up in. There are also questions about the importance of psychological continuity – for instance, if I have a serious brain injury that means I lose my memory and develop different characteristics, am I still the same person? We also know that genetically identical twins have separate identities even though their genetic identity is the same.

Against this questioning of traditionally accepted genetic accounts of identity, alternatives have been put forward. For instance, it has been suggested we could use what has been called a 'place-holder' idea of personal identity when it comes to considering whether someone is harmed or wronged by a particular choice or action.[42] Under this notion of personal identity, we might talk about 'Becki and Bob's first child' rather than consider the particular timing and thus the particular genetic make-up of this child. The claim here is that if I make a choice that means that the life of my first child goes worse than it might have gone, then I have done something wrong to this child and they can complain that I have not considered their interests in a way that is ethically responsible.

This way of conceiving personal identity does allow us to understand why we might find the conclusion to the Non-Identity Problem uncomfortable. On this view of personal identity, the Non-Identity Problem is a same person choice, and if we choose to implant a deaf IVF embryo rather than a hearing IVF embryo, the child we have created can be said to have been harmed by this choice. This is because 'the child we have created' is deaf rather than hearing, and we take the identity of this child to be the 'child we have created' in this instance rather than the child who results from a particular genetic blueprint.

While this conception of personal identity fits with many of our intuitions about the morality of attempting to avoid the creation of disabled children, I argue that this conception is one that is highly counter-intuitive on other levels. While we may accept that personal identity cannot be reduced to only genetic identity, it does seem that our genetic make-up is an important part of this identity and an important part of the psychological continuity that is so crucial to our sense of identity. Imagine that my parents conceived a year after the time they actually did conceive me (their second child). As a result, their second child was created with a different egg and sperm. It seems difficult to make sense of the claim that their second child was still me. This seems even more difficult to accept given that if I had been conceived a year earlier this further child of my parents could have existed (their third child) as well as me.

Further, at least some forms of this anti-essentialist argument rely on the notion that being born as a disabled person is necessarily worse than being born without this disability. On this view, it would be argued that we can understand that the person who is born from a particular reproductive choice is worse off than they could have been if they were born without a condition like deafness or Down syndrome, which we consider to be conditions with necessarily negative consequences. However, it is not clear that deafness or Down syndrome or other conditions that are usually seen as disabling are necessarily negative in a way that makes a life go worse for the person with that condition. Whether this is the case will depend on the particular expression of this condition, the temperament of the person and all the other factors that go to determine our quality of life, including personal and social support. While there may be some rare and extreme conditions that are extremely likely to make a life go worse than a life without that condition, in the main, the sorts of conditions we are concerned with in this book – deafness, Down syndrome, achondroplasia, being born to a disabled parent and so on – are not conditions that will necessarily make a life go worse than another life if we are to take the testimony of those living with these conditions seriously.

## *Wronging*

Other arguments that aim to solve the Non-Identity problem argue that we wrong the person who is born with a disability. The argument here is that choosing to bring to birth someone with a disability rather than someone without a disability is to do something wrong to the person who experiences this life. This is often a rights-based argument where those who take this position argue a 'child has a right to be born into good enough circumstances'[43] or with the 'right to an open future'[44] or with what might be called a 'decent minimum standard'[45] of life. The idea is that it is not enough to aim to ensure that the children we bring to birth have intrinsically valuable lives. We have a moral imperative, it is argued, to ensure that they meet a higher welfare threshold; if they do not, we have infringed their rights and thus have wronged these children. Being born to parents who may face more challenges than others when raising a child, or with a disability may mean they do not meet this 'decent minimum standard' of welfare. Thus, we do something wrong to them by allowing them to be born in these circumstances.

While this position might allow us to retain the idea of a moral obligation to bring to birth the 'best' child possible, it is not easy to defend. It is difficult to understand why we wrong someone if we accept that their life will likely be one they find intrinsically valuable. As we have seen, if we think that being alive is generally a good thing despite the inevitable negative aspects of life, then it is difficult to understand why we need a higher threshold than this idea of an intrinsically valuable life. If someone is likely to value their life and think that it is one they are happy to live and wish to continue living, then arguing that they have been wronged by experiencing this is difficult to justify.

Further, deciding where this threshold lies will be very difficult. David Archard has suggested that this threshold would mean that a minimum standard of acceptable welfare would be a life 'in which the child has the reasonable prospect of enjoying a good number of those rights possessed by all children' as outlined by, for instance, the United Nations Convention on the Rights of the Child.[46] But what is a 'good number' here, and are some rights more important than others, and how do we decide?

As we will explore in some detail in the next chapter, bias and other factors mean that we are not very good at assessing the quality of other people's lives, particularly when those lives are ones that we would consider disabled. It could be argued that trying to decide what is a threshold for a minimally decent life is highly subjective and perhaps highly susceptible to bias. Again, it seems difficult to reconcile this argument with the testimony of those living with deafness or conditions like Down syndrome or with parents who might have more challenges than others. Research seems to show that these individuals are as likely as anyone else to value their lives.

Further, there will be many individuals who appear to meet this threshold of a minimally decent life but, for whatever reasons, do not value their life in the way that others do. Trying to determine with any accuracy before conception or implantation the likely quality of life of a human being will be fraught with difficulties. We know that poverty is one of the strongest indicators of a lack of physical and mental well-being.[47] If we are required to ensure that children meet this higher standard, not only of an intrinsically valuable life but of a high threshold of 'minimally decent' life, one way to achieve this with some accuracy might be to prevent, where possible, those in poverty from reproducing. If we really are concerned with reaching a higher welfare threshold, then simply focusing on the perhaps more socially acceptable targets of disability could not be justified – we would need to consider factors that might preclude welfare more generally.

Further, as we will explore later, any position that measures anything other than the welfare of individual persons can be difficult to defend. The idea of a minimally decent threshold for the welfare of future lives seems to focus not on the expected lived experience of those brought to birth, as this threshold is higher than the threshold between what might be thought of as intrinsically valuable and intrinsically harmful lives. A threshold that infers that those with intrinsically valuable lives but lives below what others see to be as minimally decent, appears to impose an external measure of what is an acceptable quality of life. As we will see later any external judgements of this kind are difficult to justify.

## Tangible harm caused by restricting reproductive choice

While coming up with a compelling account of the harm prevented by curtailing reproductive choice is highly problematic, identifying the harm done by regulation that restricts reproductive choice in the way we have seen is, by contrast, very easy.

There are individuals and couples who, like many other prospective parents who undergo IVF treatment, wish to have a child who is 'like them' and so might want to have a child with the same disability they have, like deafness or achondroplasia, or at least not actively avoid this chance, but are prevented from doing so by regulation. Some individuals have felt coerced into prenatal screening and even, in some cases, felt pressured or even coerced into a termination as a result of Down syndrome being diagnosed in pregnancy.[48] Those attempting to access fertility treatment with diagnosed conditions like a learning disability, autism and other challenges may find getting the same access to treatment highly problematic when they simply want to parent a child as they would have been able to do if they happened to be naturally fertile. More generally, there are a great many individuals and families living with Downs syndrome, deafness and many other conditions, or with a parent who faces challenges, who find the existence and wide acceptance of policies that restrict reproductive choices based on concerns about the welfare of children like them or their children not only offensive but also harmful. For these individuals and communities, these state-sanctioned regulations, even if they do not directly restrict their choices, send a clear message that avoiding the birth of someone like them or their loved one is something we should all wish to avoid. This perpetuates and reinforces existing discriminatory attitudes towards those with these conditions and fuels other discriminatory behaviours and practices. These are real and tangible harms that affect existing people. So even if we could make sense of impersonal harm that affects a population rather than an individual in this context, when weighed against the small number of choices that these restrictions apply to and the tangible harms to actual people that these restrictions directly create, it seems clear that these person-affecting harms must outweigh these other problematic considerations.

## Conclusion

The conclusion that there may not be any good reason why we should interfere with choices to bring to birth a disabled or disadvantaged child feels highly counter-intuitive to many of us. Parfit called this the Non-Identity Problem as we appear to have an argument that justifies choices to bring to birth lives that we may find problematic. In this chapter, I have outlined the main attempts to provide ways to retain the intuition that choosing to bring to birth a disabled or disadvantaged child is ethically unjustified. However, none of these attempts to solve the Non-Identity Problem appear to provide the robust reasons we would need to justify interfering with the reproductive choices of others and reinforcing negative social norms around disability.

# CHAPTER 5
# SHOULD WE ACCEPT THE CONCLUSIONS OF THE NON-IDENTITY PROBLEM?

As we have seen, the Non-Identity Problem is a problem, not an argument, for many because they find accepting the idea that it might be ethically acceptable to choose to bring to birth a disabled life highly counter-intuitive. In this chapter, I will argue that we can provide convincing reasons that explain this intuitive unease to the conclusion of the Non-Identity Problem. If I am right about this, then these feelings of unease are simply feelings and do not provide good reason to justify an obligation to bring to birth the 'best' child possible. If this is the case, then there seems nothing in the way of us accepting the conclusion of the Non-Identity Problem, that so long as a child is likely to have an intrinsically valuable life, we do not do harm or wrong by bringing them to birth.

## Rational preference, common-sense morality or intuition?

To remind ourselves, Harris and Savulescu, both promote this idea of a moral obligation to bring to birth the 'best' child possible. The main way that they defend this idea is by demonstrating the intuitive unease they feel about the Non-Identity Problem by following Parfit's lead in providing what might be called Non-Identity examples. These examples involve hypothetical choices we may have between choosing to bring to birth individuals who will have a condition seen to be disabling or face other challenges, and the alternative of bringing to birth lives that we assume do not have these foreseeable challenges or conditions. The idea that we might choose to bring to birth someone who we think will have a 'worse' life than another possible life produces a feeling of unease in many of us. The use of these examples attempts to show that because we *feel* that a choice is wrong, it, therefore, must actually *be* wrong to make these choices. For instance,

Savulescu introduces a Parfit-inspired example of a woman deciding to delay conception until after her rubella infection in order to attempt to avoid having a child affected adversely by this infection and concludes simply, 'She should choose to wait until her rubella is passed'.[1] Similarly, for Harris, it is just clear that a prospective mother 'has reason to do what she can to ensure that the individual she chooses is as good an individual as she can make it […] and that will have the best possible chance of a long and healthy life and the best possible chance of contributing positively to the world it will inhabit'[2] even if he does not provide clear reasons why this might be so.

Savulescu and Harris both use examples of choosing IVF embryos to be implanted. In Savulescu's example, one embryo has no 'abnormalities', but the other shows a predisposition to asthma.[3] Harris' example involved a choice between six IVF embryos, three of which are shown to have 'various genetic disorders and three seem healthy'.[4] These examples are used to elicit an intuitive response from us. Many of us respond to examples where a choice is made to bring to birth a child with conditions that are seen as disabling with feelings of unease. Harris argues that these feelings of unease indicate that we have a 'rational preference'[5] to choose to bring to birth the best child possible, and Savulescu argues that the unimpaired embryo 'should (on pain of irrationality) be implanted'.[6] They claim that choosing the 'unimpaired' embryo or 'unaffected' child is the rational choice and, thus, the right choice. However, without explaining why this choice is rational and providing arguments to back up their statements, I argue, they rely on our intuitive reactions of unease in response to these examples alone, rather providing reasons to explain the wrongness of this choice.[7]

## The role of bias and unconscious thinking in intuition

Earlier in this book, we explored the role of intuition in ethical decision-making. I argued that our initial response to an ethical question may be one that is influenced significantly by intuition and, thus, unconscious thought processes. Therefore, this initial response may not align with what we might consider to be our core values and may, unwittingly, be influenced by unconscious bias. As a result, I argued that we should use something like the ARC approach to develop an answer to ethical questions that is more than our initial, quick answer but an answer we have deliberated about and provided reasons for and defended against possible counterarguments.

Thus, while we may have a strongly negative intuitive response to the conclusion of the Non-Identity Problem and the sort of Non-Identity Examples that are put forward by Parfit, Harris and Savulescu, this is not enough to justify rejecting its conclusion. If we want to reject the conclusion of the Non-Identity Problem and attempt to establish a moral obligation to bring to birth the 'best' child possible, then we need to do more than explain that this feels odd or uncomfortable – we must be able to give reasons to support this claim and show how it can be defended against counterarguments. But while there have been attempts to provide these reasons to support this idea of a moral obligation to bring to birth the 'best' child possible, as we have seen, these quickly run into their own issues, and a convincing solution to the Non-Identity Problem has been elusive.

Interestingly Harris and Savulescu do not attempt to offer a solution to the Non-Identity Problem. They hint at the idea that the existence of some kind of impersonal harm is the cause of our intuitive discomfort here by talking about a choice to bring to birth a disabled or disadvantaged child life, making the 'world a worse place than it need have been'[8] or appealing to 'harmless wrong-doing'.[9] But, they do not explore this issue in any detail or attempt to provide any reasons why we might accept this unusual conception of harm or defend it against the many counterarguments we have seen that can be directed at this concept. In fact, Savulescu (with Kahane), when writing about the nature of the possible harm that might be caused by choosing to bring to birth a child who they consider to have a life with more challenges than another alternative life say that '[w]e do not take a stand on this difficult philosophical issue. As we have tried to show, our moral intuitions about timing of conception recognize reasons to select future children'.[10] Given the difficulties in developing a convincing account of impersonal harm, Harris and Savulescu fall back on defending their arguments mainly through the intuitive feelings of unease that are produced when they present Non-Identity Examples. But as we have seen, while intuition may, in some instances, indicate strong ethical positions, this cannot be taken for granted. Much more work needs to be done to justify a claim that we have a moral obligation to bring to birth the 'best' child possible.

I argue that rather than having to try and prop up complicated and problematic notions of impersonal harm or alternative notions of personal identity or harm, we can work to understand why this intuitive feeling of unease occurs in response to Non-Identity Examples. In doing this,

I argue we can show why these intuitive feelings are simply feelings and not indications of good reasons in these particular cases.

I suggest that by understanding the unconscious thought processes that contribute to these intuitive reactions, we can gain insight into why we have these intuitive responses to Non-Identity Examples. We know that a great deal of our cognitive activity is unconscious and that these unconscious processes draw on cultural norms and stereotypes in order to allow us to make quick decisions and assess situations and people quickly. This results in unconsciously biased thinking around stereotypes of all sorts of characteristics, including, for instance, what it means to be gay, female, black or disabled.

Jean Moule explains the way that unconscious bias works in terms of race, saying that 'ethnic and racial stereotypes are learned as part of normal socialization and are consistent among many populations and across time [...] when we receive evidence that confronts our deeply held and often unrecognized biases, the human brain usually finds ways to return to the stereotypes'.[11] These biases and stereotypes are often reinforced by the media; thus, even if we do our utmost to guard against racial bias, we are unlikely to succeed. As Moule continues, '[i]t is important to note that the *well intentioned* are still racist'.[12] He considers the dominant norms and standards that pervade our daily life around issues of race and gender and quotes Barbara Applebaum, who argues that

> Because many people believe these norms and standards are culturally neutral and universally right, true, and good, they do not understand how these norms and standards oppress others. They are not even aware of this possibility and, in this sense, such racism is unintentional.[13]

While Moule and Applebaum primarily address unconscious bias as it applies to race, it is easy to see why it might be important to explore our often-shared attitudes to disability.

## The unconscious, bias and ableism

Ableism, like racism and sexism, 'describes discrimination towards a social group, in this case disabled people'.[14] But while it is natural to compare ableism with racism and sexism, these comparisons do not capture the extent of ableism and its internalization in society generally. There are those among us who display both conscious and unconscious racism and sexism. Others of us are consciously, or at least publicly, committed to equality but display unconscious bias regarding race and gender. However, it is thought that the way the conscious and unconscious thought processes work with ableism is significantly different.

It has been suggested that 'the majority of people hold unconscious prejudice towards disabled people despite consciously having low levels of prejudice'.[15] As a result, 'nondisabled people may believe they feel positively towards disabled people but actually hold negative attitudes which they disassociate or rationalize.[16]' This mismatch between our conscious and unconscious thought processes means that 'nondisabled people may try to appear sympathetic and supportive of people with disabilities, but they may show signs of anxiety (e.g., averted gaze, closed posture, greater interpersonal distance) that reflect their implicit attitudes without personal awareness of what they are communicating nonverbally.[17]' These implicit or unconscious attitudes towards disabled people mean that non-disabled people often distance themselves from disabled people and, in an attempt to be sympathetic and supportive, can act in ways that are perceived as patronizing and infantilizing by disabled people.[18] It is suggested that ableist discrimination associated with unconscious bias tends to be conveyed subtly. This might involve avoiding interactions or close contact with disabled people and manifesting anxiety where interactions cannot be avoided.[19]

Disability appears to elicit unconscious feelings of disgust, fear and anxiety in the majority of non-disabled people.[20] These feelings can be explained in a number of ways. As with other unconscious biases, we tend to feel comfortable and favour those who are like us and conversely disfavour and distance ourselves from those who appear to belong to a different group from us. However, it is not just 'otherness' that produces these negative reactions in many of us. It is thought that physical disability may create a primal sense of fear in us in the same way as snakes and spiders do for many of us. The suggestion is that physical signs of disability may have associations with danger of disease and contagion. Further, it is thought that physical disability causes discomfort as it arouses fear of death and possible physical decline.[21]

One of the reasons that ableism is thought to be so widespread is that '[i]t may be that destructive disability portrayals, representations, and stereotypes are so prominent that they are commonly accepted and not viewed as negative. As such, most people are probably not conscious of the ways their understandings are problematic'.[22] Strongly negative notions of what it means to be disabled are cultural norms in our society and thus may not be questioned. This is clear in the dominant attitudes to bringing to birth a child with a condition seen as disabling. When people are pregnant, they often say things like, 'I don't care whether they are a boy or a girl as long as they are healthy'. This idea of 'healthy' usually includes the idea of not having a condition that is viewed as a disability even if this condition is still compatible with an intrinsically valuable life. Those who do have a child with conditions such as deafness, Down Syndrome, autism or achondroplasia, or those who are close to someone with these conditions or characteristics may change their feelings about these conditions. In fact, we know that one of the most successful ways of changing our negative unconscious attitudes towards disability is to have more contact with disabled individuals, as this helps to overturn negative stereotypes.[23] However, without this experience of knowing someone living with conditions seen to be disabling, stereotypes, cultural norms and negative messaging about disability means that the idea of having a child with one of these conditions can seem to be unquestionably a terrible thing to happen because this is what we have been led to believe by general entrenched attitudes and perceptions of disability.

The routinization of screening for conditions like Down syndrome in pregnancy reinforces and perhaps even amplifies the existing negative social norms around the condition. For instance, in 2010 Dena Davis explained '[t]wenty years ago, seeing a woman in the supermarket with a child who has Down syndrome, my immediate reactions were sympathy and a sense that that woman could be me. Now that testing for Down syndrome is virtually universal in the United States, when I see such a mother and child I am more likely to wonder why she didn't get tested'.[24] It seems we have, at least to some extent, internalized the view that not only should parents of children with conditions like Down syndrome be pitied, but also that we should condemn them for what we see as their choice to have a child with this condition. While having a child with Down syndrome may be a tragedy for some parents, and I absolutely uphold the right of any person to terminate any pregnancy for whatever reasons are important to them, this assumption that having a child with Down syndrome is a tragedy is certainly not the experience of many families who have a child with this condition.

Negative attitudes to disability are commonly accepted in our society and reinforced by regulations such as routine screening in pregnancy. Many of us have strong feelings that we would prefer not to have a child with a condition seen as disabling, and we feel uneasy about creating human lives that may face more challenges than others. Given these usually unquestioned social norms and our strong unconscious reactions to disability, it may seem to many that it is just 'commonsense morality'[25] or even a 'rational preference'[26] to support regulations reinforcing these feelings. As a result, compassionate and generally very well-intentioned individuals may hold the view that we should avoid the creation of impaired lives wherever possible, and they would likely be horrified to be considered biased or prejudiced against those who live with disabling conditions or in challenging circumstances. But like other unconscious biases, it seems that this negative attitude to disability and disabled people may not come from the place of compassion and sympathy that we think it does but from unconscious, primal and irrational fears of 'otherness', disease and death.

## The disability paradox

Further, perhaps because of these irrational and unconscious fears about disability, as non-disabled people, it seems we are very bad at evaluating the quality of life of those living with a disability. This is known as the 'disability paradox'[27] and is explained by Tom Shakespeare:

> Have you ever thought to yourself: 'I'd rather be dead than disabled?' It's not an unusual reflection. Disability, in everyday thought, is associated with failure, with dependency and with not being able to do things. We feel sorry for disabled people, because we imagine it must be miserable to be disabled. But in fact we're wrong.[28]

The paradox here is that while many non-disabled individuals view those living with disability as people with lesser welfare than themselves, when we ask those living with a wide range of disabling conditions, they consistently report high levels of quality of life, which is often equivalent to or even higher than the quality of life reported by those without these conditions.[29] There have been suggestions that there is a problem with this self-reporting here and that those with disabilities are not reporting accurately.[30] However, investigation of this claim has concluded that 'to date, across a wide range of studies, the best available evidence suggests that such self-reports are largely accurate'.[31] As a result, suggesting that disabled people are somehow in denial is, as Shakespeare argues, 'patronising' and 'insulting'.[32]

While there will be some conditions or illnesses that entail a huge amount of suffering, to the extent that the person living this life may well give a very low quality of life score, it seems that these conditions are rare, and this is not usually the situation for those living with the sort of disabilities that are typically the focus of these ethical debates – deafness, Down Syndrome, achondroplasia, and so on. What is clear is that it is very difficult to make an accurate assessment of someone else's quality of life, and the evidence is that physical disability may not make a significant difference to an individual's quality of life. What may make more difference is how society accepts all individuals and modifies our environment to allow everyone to participate as fully as possible.

As a result, we must be prepared to scrutinize decisions based on these assumptions around welfare and disability much more carefully and be ready to accept the biases that we know skew these assessments. We must be willing to consider that while these are widely held views, they may not necessarily reflect the reality of living with a disability and may unintentionally oppress others. If we really are well intentioned, then we must at least seriously investigate this possibility.


## Everyday conception of harm as a comparative thing

There might also be other things, apart from unconscious bias and inaccuracies in assessing others' quality of life, that feed into our intuition about Non-Identity Examples. I suggest that another reason we find the conclusion of the Non-Identity Problem difficult to accept is that we tend to apply our usual concept of comparative harm and benefit to these cases. However, as I will explain, while our everyday notions of comparative harm and benefit work well with decisions about existing people, they do not make sense when the subject of this harm or benefit does not yet exist.

In our everyday contexts, when we consider the effect of an action or choice on the welfare of an existing individual, we talk about whether someone is harmed or benefitted by an action or a choice. Our everyday use of the words 'harm' or 'benefit' and our understanding of these concepts are comparative notions. We think about a harm as something that makes people worse off than they would otherwise be and a benefit as something that makes someone better off than they would otherwise have been.

When it comes to existing children, the choices we make will affect the welfare of this particular future individual. In these instances, where our choices will make a difference to the welfare of these existing children, our usual comparative notions of harm and benefit will be applicable when assessing these choices. In these instances, where we can harm or benefit the welfare of specific existing individuals, we can compare their lives with or without the intervention we are considering. For example, removing an existing child from one family to live with another will impact the quality of this child's life for better or worse. We can weigh comparatively whether their welfare is likely to be increased or decreased. Similarly, if we were to deafen a hearing child, this would affect this particular child's welfare. Thus, these comparative questions about whether this child will be worse off or better off, given this action, are applicable here.

When it comes to choices and actions that will affect the welfare of particular existing individuals, then comparative questions are appropriate. For example, we might ask, 'Will this child be likely to be better or worse off if we

- Provide a specific treatment?
- Change their living arrangements?
- Provide specialist support?

Because we take this comparative approach to consider the welfare of existing children, we tend to apply this same approach to decisions about whether to bring *future* children to birth. But when the child we are concerned about is a future child, one that does not yet exist, these comparative notions of harm and benefit no longer apply.[33]

Unlike cases of considering the effects of an action on the welfare of an existing individual, where we can ask whether someone would be worse or better off if an action is taken, these preconception and pre-birth cases are very different. The actions we are considering in these cases are not ones that will affect the welfare of an existing person; they are actions that will cause a particular person to come into existence. For instance, in the case of access to IVF, we are not asking whether an existing child should live with particular parents or whether we should place them with other parents, as is the case with adoption, but whether we should act to enable a child to be born in the only conditions they can be, to these particular parents at this specific time.

Because we are deciding whether someone should be brought into existence, these comparative notions of harm and benefit just do not work, as there are no instances of welfare we can compare here. The only comparison we can make is between nothing (non-existence) and the welfare of a person who may come to exist. Having nothing to compare with something means that our usual comparative notions of harm and benefit and our usual assessment of welfare do not fit these more unusual examples, as there is nothing to compare this existence to.

As we saw earlier, for many reasons, our minds do not deal with the notion of non-existence very well. This may be because there is literally nothing to imagine, and we find that difficult. It may also be that the thought of our own non-existence, either before conception or after death, is such a difficult thing to contemplate. But, I argue, the main problem we have in these cases is that we are so used to making decisions in these comparative ways that our minds try and find a comparison to make where there is none.

We see this when we watch those TV programmes where someone investigates their family tree. Often, people on these programmes say something like, 'Wow, if my great, great grandfather hadn't travelled to England from Ireland, I'd have been Irish'. Here, this person is doing what they are used to doing in everyday life and are trying to compare the effect on themselves of their great, great grandfather's action or inaction. They automatically think of two states of their own existence that can be compared depending on what decision their ancestor took. They imagine being born in England or, in the case of their relative staying put, in Ireland. The reality here is that this person would not have been born at all if this move had not occurred or the millions of other actions needed to take place to enable their birth. The more appropriate comparison to make here is between being born in England as they were and not being born at all. But perhaps because we are so used to thinking in these comparative ways, or perhaps because we find the thought of our own non-existence so difficult, we tend to automatically revert to imagining a comparison that, on reflection, cannot be made.

My suggestion is that our tendency to revert to thinking in terms of comparisons may explain further why we feel uncomfortable about a choice to bring to birth a child who will be deaf, have Down syndrome or be born into what we see as less than optimal circumstances. In these cases, I suggest that there is a tendency to ask questions like:

- Would this person be better off if they weren't deaf or didn't have Down syndrome?

- Would it be better for this child to be born to parents without these challenges?

- Would I want to be deaf or born to parents with particular challenges?

While it may be natural to ask these questions when faced with these choices, the comparisons that these questions attempt to make are not comparisons that are possible. The deaf embryo can only become a deaf child, not a hearing one. The child born with Down syndrome could not have been born without this condition. The child born to a mother with a mental health issue could not have been born to another mother.

## Welfare-affecting and non-welfare-affecting choices

We can understand the issue a bit more here if we think about this in terms of welfare-affecting and non-welfare-affecting actions or choices. When we make choices about how to act with regard to existing people, there are two different states these existing people could end up in depending on the actions taken. As a result, in the case of existing individuals, these are welfare-affecting choices. We can make these existing people better or worse off by our choices and our actions. Thus, applying a comparative notion of harm and benefit makes sense in these cases as it makes sense to say, 'Would X be better off if I do make this choice or another choice?'.

However, when dealing with choices that decide whether a future child will be born, these are non-welfare-affecting choices. The child we are considering bringing to birth will either be born with the condition or in the conditions they can be born in, or they will not be born at all. In these cases, our usual comparative notions of harm and benefit do not make sense as we are trying to compare something (existence in the only conditions this child can be born in) with nothing (non-existence).

Acknowledging and understanding this distinction between choices and actions that affect the welfare of a particular child and those that do not but affect who will be born helps us understand why we are uncomfortable with the conclusions of the Non-Identity Problem. I suggest that part of the strength of the intuition we have that choosing to bring to birth a disabled child is wrong arises from blurring this very important distinction between choices that harm someone's welfare and choices that do not.

To explore this further, let's look at an example provided by Harris. Harris argues that while a deaf child has 'no complaint because for them the alternative is non-existence',[34] we still do something wrong in allowing a child to be born in what he calls a 'harmed' state, a state that anyone might have a rational preference not to be in.[35] Harris argues that existing with disabilities such as deafness is clearly 'worse' than existing without these disabilities,[36] even though he believes that the child has benefitted by being brought to birth overall.[37] To explain the nature of this wrongdoing, Harris compares choosing to bring to birth a deaf child when a hearing one could be chosen with failing to provide an available cure for deafness or deafening a hearing child. He argues, 'I do not believe there is a difference between *choosing* preimplantation deaf embryo and refusing to cure a newborn. Nor do I see an important difference between refusing a cure and deliberately deafening a child'.[38]

For Harris, the moral wrong done by bringing to birth a deaf child is a serious moral wrong, as wrong as deafening a hearing child. If choosing to bring to birth a deaf child was really morally equivalent to deafening a hearing child or preventing a deaf person from accessing treatment that would allow them to hear, then we can understand why this choice and similar choices that unnecessarily negatively impact the welfare of people make us uncomfortable. Deafening a hearing person or preventing a deaf person from accessing treatment that would allow them to hear would be to damage that person's welfare. Harris argues that the reason that this is the case is that deafness itself is a harm and that choosing to bring to birth a deaf child is also to do harm.

While this explanation from Harris is initially appealing, there are two serious problems with this analogy between choosing to bring to birth a deaf child and deafening a hearing child/prohibiting a deaf individual access to treatment that would allow them to hear.

The most fundamental problem here is that, while initially seeming similar, these cases are significantly different. In the case of choosing to bring to birth a deaf child, either by choosing a 'deaf' IVF embryo or continuing with a pregnancy where the fetus has been identified as deaf, we do not negatively impact the welfare of any individual by allowing the embryo to be implanted and brought to birth or by continuing an established pregnancy. In these cases, the deaf child is either born in the only condition they can be born in or not born at all. Our choice to bring them to birth does not change the conditions this child experiences when they are born.

The cases of deafening a hearing child or preventing a deaf person from accessing treatment they wish to have that would allow them to hear are very different. In these cases, the proposed actions will negatively impact those existing people. In these cases, these individuals can exist as hearing or deaf people; these options are open to them, and thus, our actions can make these individuals worse or better off.

While all three actions, bringing to birth a deaf child, deafening a hearing child or denying a deaf individual treatment to allow them to hear, result in the existence of a deaf person, these are very different actions and choices. Making someone worse off than they otherwise could have been is something that we would all wish to avoid. However, bringing someone to birth in the only condition they can be and with a life they are as likely as anyone else to value does not make anyone worse off than they otherwise could have been. This particular person can only be born in the condition/s they are born in or not at all.

I argue, unlike Harris, that being deaf is not necessarily a harm. We know from the lived experience of deaf people that they are as likely as anyone else to have lives of quality and well-being that they value.[39] The harm involved in deafening a child or denying treatment that would allow someone to hear is the thwarting of individual choices and denying an individual something they might value. We would not pressure a deaf adult to accept treatment that might allow them to hear, even if we feel that not hearing is a disability or something we would not want for ourselves. However, because the majority of people would probably prefer not to be deaf, we would feel it important to avoid deliberately deafening individuals and to give deaf individuals who wish to access treatment to allow them to hear the opportunity to do so.

Despite these cases not being analogous, this analogy does help us to understand the unease many of us feel about a choice to bring to birth a child who may have more challenges than other possible children. We instinctively want to avoid harming people, and, in some cases, making someone deaf is to harm them as it may make them worse off than they might otherwise have been. However, while this is true in some cases of deafness, where options that would otherwise have been available to someone are denied, this is not the case when we consider bringing to birth a congenitally deaf child. If we are not clear about this distinction, it is easy to see why many of us, like Harris, may initially see making a deliberate choice to bring to birth a deaf child as analogous to the deafening of a child and be very concerned about this choice.

We may have good reasons, based on harm prevention, to intervene when it comes to welfare-affecting choices. For instance, it is important that we are very careful about where we place existing children for adoption and should try to provide them with the best family environment for them we can. It may be justified to intervene with choices around other existing children when it comes to deciding what the most appropriate medical treatment or social support is for these children in order to protect them from harm and maximize their welfare. It may even be justified to put some pressure on pregnant people to accept testing for infections like HIV and syphilis, given that doing so may prevent harm to this particular fetus and, thus, the future child they will become.[40] These are all instances where we can compare the outcomes for a particular child depending on the action taken and assess how to act to increase their welfare. However, in the cases we focus on in this book, where we are contemplating the welfare of future children that do not yet exist, this is not the case. These children can only exist in the condition they can exist in, as being deaf, having Down syndrome, or being born to parents who may face their own challenges or not at all. Choosing differently will not allow these children to be born without these challenges. It will only result in them not being born at all. While our minds might blur these kinds of welfare-affecting and non-welfare-affecting cases together and result in a feeling that we should protect someone from what we see as harm, it is important here that we recognize that in the non-welfare-affecting cases, no harm is avoided by preventing the birth of lives that might be considered to be impaired but are likely to be intrinsically valuable to those who experience them.

## Harm to others

A final reason that might also contribute to our unease about accepting the conclusions of the Non-Identity Problem is to do not with the welfare of future children, but with possible harm to existing people. If we think about Parfit's fourteen-year-old girl example, it seems that having a child at fourteen, while it is unlikely to create a child who has a life that they do not find intrinsically valuable, is likely to impact that fourteen-year-old girl severely and negatively. In most cases, a fourteen-year-old experiencing pregnancy and childbirth and either adoption or parenthood at this very young age is likely to cause real concern in those around her and those who hear of her plight. Similarly, we may be concerned for the well-being of parents who are choosing to bring to birth a child with a disability or in particularly difficult circumstances. There will be some prospective parents who may well be negatively impacted by the birth of a disabled child or the additional challenges of a child born into difficult circumstances, and our feeling of unease might be a response to this. It is also important to recognize that, given the shared overwhelming negative attitudes that many of us have toward disability and our tendency to assess the quality of life for those with disabilities more critically than other lives, we might regret the birth of a child with these conditions because of the impact we may *believe* this birth will have on the lives of the family that this child is born into even if that negative effect is not felt by their family. As a result, our intuition here when it comes to examples like the fourteen-year-old girl or another individual who, for instance, chooses to bring to birth a deaf child rather than a hearing child may, at least in part, be due to the concern we have for the parents in these scenarios. We might be concerned that these individuals are making their own lives harder than they need to be and that these choices will harm them as individuals.

However, if respecting the choices of individuals is important, then it seems that empowering individuals to make the right choices for themselves is important here. Of course, there will be individuals who choose not to have a child in particular circumstances or choose not to have a child who may have a particular condition, and this may well be the right choice for them. However, there will be others who wish to make a different choice for whatever reasons they find important. If we believe that respecting autonomy and, thus, the choices that individuals make is important, then we should respect *all* reproductive choices, not just the ones that align with the dominant social norms of our society. As we have seen, these social norms have arisen through bias, stereotypes, unconscious thought processes and fear and are reinforced by established regulations which may make them seem beyond reproach. Where there is no compelling argument that the resultant child will be harmed by a choice to bring them to birth, whatever our persisting concerns for the child and the potential parent, we should empower individuals to make the choice that is right for them. While it may be natural to feel protective towards future children and future parents, and this may contribute to our intuitive feelings of unease in Non-Identity Problem examples, where there is no compelling evidence of harm to this resultant child, these feelings are misguided. In fact, harm might be caused by not allowing some individuals to make the reproductive choices that are important to them or by regulation that seems to put a lower value on their lives and the lives of their loved ones.

## Contrary preference thesis

As we have seen, there are lots of explanations of what might contribute to our intuitive reaction of unease to cases where individuals choose to bring to birth disabled children or in conditions we feel are disadvantageous. These choices go against strongly held and deeply embedded social norms.

Misplaced protective concerns for the welfare of children and prospective parents may also contribute to concern about these choices.

But in the absence of any compelling account of the harm that is done by allowing these choices, I argue that these intuitive reactions to choices to bring to birth children in what others consider to be difficult circumstances should be regarded as feelings or preferences that cannot be imbued with any moral weight.[41] While many of us may wish to implant one IVF embryo rather than another, I argue that the reasons we have for these choices are not moral reasons but merely preferences about what sort of children we would like to have and perhaps what kind of society we would like to live in. Of course, these preferences are probably highly influenced by the cultural norms of our society, and as we have seen, these norms do not necessarily reflect the reality of living with conditions that are considered disabilities.

As long as we create lives that we expect to be intrinsically valuable, we do not negatively impact anyone's welfare. Thus, I argue, it is a matter of moral indifference which lives we choose to bring to birth. Of course, many of us have a preference not to choose to bring to birth a disabled child and, just as we find other preferences that might contradict cultural norms difficult to accept or even unsavoury, we may find a preference for a child with a condition usually regarded as a disability difficult to accept. But as long as we are choosing to create lives we expect to be intrinsically valuable, I argue, these are simply preferences for the sort of children we would like to have or preferences about the sort of world we would prefer to live in, and there can be no moral obligation to choose one life we expect to be intrinsically valuable over another.

If we wish to argue that these choices indicate something morally or ethically relevant, we need to do more than just say that they make us feel uncomfortable. There are many things that make us feel uncomfortable now and in the past. Many people felt uncomfortable about racial and gender equality or the acceptance of homosexuality or interracial marriage, but without good reasons to oppose those who act positively towards these issues, we are simply expressing our own personal dislike or, in some cases, prejudice.

While I argue that these preferences are not subject to any moral obligation to choose one way or another, they are important. In order to respect individuals' autonomy and allow us to have the sort of control over our lives that we think makes our lives more fulfilling, in the absence of moral reasons to interfere, we should allow individuals to make the reproductive choices that are right for them. In the same way, we allow individuals to make their own choices about whether to have medical treatment, who to spend their lives with and how to live their lives (so long as no harm is caused to others by this choice), we should also respect the reproductive choices of prospective parents. For instance, most of us choose a sexual or companionship partner based on preferences or attraction. These companions may become reproductive partners as well as social partners. If we choose a reproductive partner who has particular racial, aesthetic, intellectual and temperamental traits that may be inherited by any genetically related children, as long as the resulting child's life is likely to be intrinsically valuable to that child, then choosing to reproduce with this particular partner rather than another is merely a matter of preference and thus something we should not interfere with even if we do not share this particular preference. Similarly, I argue that if a prospective parent wishes to implant a 'deaf' embryo or an embryo with any other congenital disability that is usually compatible with the notion of an intrinsically valuable life, then this is an individual preference for the sort of child they wish to have and, as long as this choice harms no one, we have no reason to override or attempt to influence this choice.

## Conclusion

The idea that we do not harm anyone by choosing to bring to birth a child who will have a disabling condition or will be born into challenging circumstances but is as likely as anyone to have an intrinsically valuable life is one that is, for many of us, counter-intuitive. In this chapter, I have explored a number of reasons why we might have this intuitive response. Our unconscious thinking appears to have a significant influence when it comes to our intuitive responses. Primal and irrational fears of 'otherness', disease and death contribute to unconscious negative attitudes towards disability and disabled people, and these negative attitudes are so widespread that they may appear to many as rational rather than discriminatory. Further, our tendency to apply comparative notions of harm and benefit to children and their parents may contribute to our intuitive feelings that someone is made worse off by choices to bring to birth disabled or disadvantaged children. However, where there is no compelling evidence of harm to this resultant child, these feelings are misguided. I have argued that by understanding these intuitive reactions, we can understand that they are simply feelings and not indications of good reasons that we need to consider in these particular cases.

# CHAPTER 6
# WHERE DO YOU DRAW THE LINE BETWEEN WHAT IS AN ACCEPTABLE AND UNACCEPTABLE QUALITY OF LIFE FOR A FUTURE INDIVIDUAL?

If we take the view that being alive is generally a good thing for most people, then it is very difficult to come to a convincing account of the harm or wrong that is done by either bringing to birth a child with disabilities like deafness, Down syndrome or achondroplasia, or a child born into other conditions that are viewed as suboptimal, such as having a parent with a disability.

As a result, I have argued that the view that regulations that restrict reproductive choices in such cases out of concern for future children are the common sense or rational approach is based largely on a complex intuitive response rather than convincing evidence that these regulations avoid harm.

While most of us are well-meaning and think and act in ways that aim to protect the interests of these future children, there are a number of cognitive peculiarities that mean that our best intentions may end up causing rather than preventing harm. I have argued that our commonly held unconscious reactions to disability mean that, often unbeknownst to us, we have unjustified negative perceptions of disability that are reinforced by social norms and regulation that also reflect these widely held views. As a result, while non-disabled individuals are often confident in their (usually negative) assessments of the quality of life of disabled individuals, this confidence is invariably misplaced.

I also suggested that another main reason we might find the conclusions of the Non-Identity Problem counter-intuitive is that our usual ways of assessing the effect of choices or actions on welfare are to make comparisons to decide if someone will be made better or worse off by these choices or actions. We intuitively think that it is 'worse' to be deaf than hearing, or 'better' to have parents with no serious challenges to looking after children, for instance.

However, by exploring these comparative notions of harm and benefit, we have identified that they are not applicable to instances where we are choosing who will be born, that is, where decisions are not welfare affecting. In such cases, children can only be born in the conditions they can be born in; thus, there is nothing to compare this/these conditions to. Thus, where we are choosing who will be born – that is, decisions that are not welfare affecting – then we need to reject questions that ask us to make these comparisons, such as:

- Would this person be better off if they weren't deaf or did not have Down syndrome?
- Would it be better for this child to be born to parents without these challenges?
- Would I want to be deaf or born to parents with particular challenges?

While these are questions that many of us feel are relevant and thus often spend a great deal of time considering when attempting to make these decisions, for the reasons I have set out, these seemingly coherent and relevant questions turn out not to represent the choice that we have in front of us when deciding about these cases.

If we accept that this comparative notion of harm and benefit does not apply in these cases, then I suggest that the more accurate but still highly problematic question we need to focus on when making these decisions is:

- Is the life that we intend to create one that is of an acceptable quality?

## Thresholds and the welfare of future individuals

Answering the question 'Is the life that we intend to create one that is of an acceptable quality?' requires us to think about thresholds between what we consider to be lives of acceptable and unacceptable quality. If we are attempting to make decisions about the welfare of future children in individual cases, we imagine the lives of these future individuals and make an assessment of whether the life they will experience is likely to be above the threshold of what we consider to be a reasonable or acceptable quality of life. Imagining these thresholds and where future lives may lie in relation to these thresholds is the only way of making decisions if we want to try and take into account the welfare of future lives.

Of course, deciding where this threshold is and how we can apply this to future lives in a meaningful way is hugely challenging. But if we hold on to the idea that it is important to protect the welfare of future individuals, then determining where to draw this threshold and why this threshold is the central question we need to address.

Any attempt to set a threshold in this way will rely on a scale against which to mark this threshold. This scale will represent the expected welfare or quality of the future life that we are attempting to assess. In Figure 2, you can see the vertical axis on a diagram that attempts to demonstrate this threshold setting. This axis represents a scale from what I have called 'Maximum welfare' to 'Minimum welfare'. The horizontal lines represent some of the many possible thresholds between what is thought to be an acceptable and unacceptable quality of life. How high on the vertical axis of welfare you put your threshold will indicate what level of welfare or quality of life you think is acceptable when it comes to creating new lives.

The first problem we face is determining what we mean by 'Maximum welfare' and 'Minimum welfare'. It may be that many of us think that we could recognize a life of high welfare or quality or low welfare or quality, but

| | |
|---|---|
| Maximum<br><br>Welfare | ? |
| | ? |
| | ? |
| | ? |
| | ? |
| | ? |
| Minimum<br><br>Welfare | ? |

**Figure 2** *Thresholds and the welfare of future individuals*

actually identifying examples of these kinds of life is more complicated than we might first think. Attempting to develop this scale of welfare or quality of life raises a huge number of questions, for instance:

- What does an extremely high quality of life look like?

- Is a high quality of life simply one that does not involve serious disability?

- What do we mean by disability?

- If we do consider those with disabilities as having a lower quality of life, how can we reconcile this with instances where those with disabilities rate the quality of their life as highly as someone who is not disabled?

- How do we measure the effects of social conditions such as poverty on welfare?

- How do we assess a life as having extremely low quality?

- We know that other characteristics such as gender, beauty and identifying as LGBTQI+ may have an impact on one's quality of life. How might this impact how we assess the quality of these lives?

We will return to these challenges in attempting to measure quality of life later, but it is important to recognize this fundamental challenge from the outset.

## Where do you draw the line?

Use the decision tree (Figure 3) to get an idea of where you might draw this line between the sorts of future lives that you see as acceptable and unacceptable to choose to create. This is of course a simplistic way to try and determine where you might consider this threshold to be, and people's actual positions will be much more nuanced than can be represented by a decision tree. However, this diagram may help you to explore the general position you take based on your answers to the questions we have been exploring.

Once you have considered where you stand on this issue, you might find the descriptions of the general positions that follow interesting. However, again, these are simplified versions of the sorts of positions people take on this issue and may not capture every position here but outline the sorts of thresholds that might be used when considering these cases of future people.

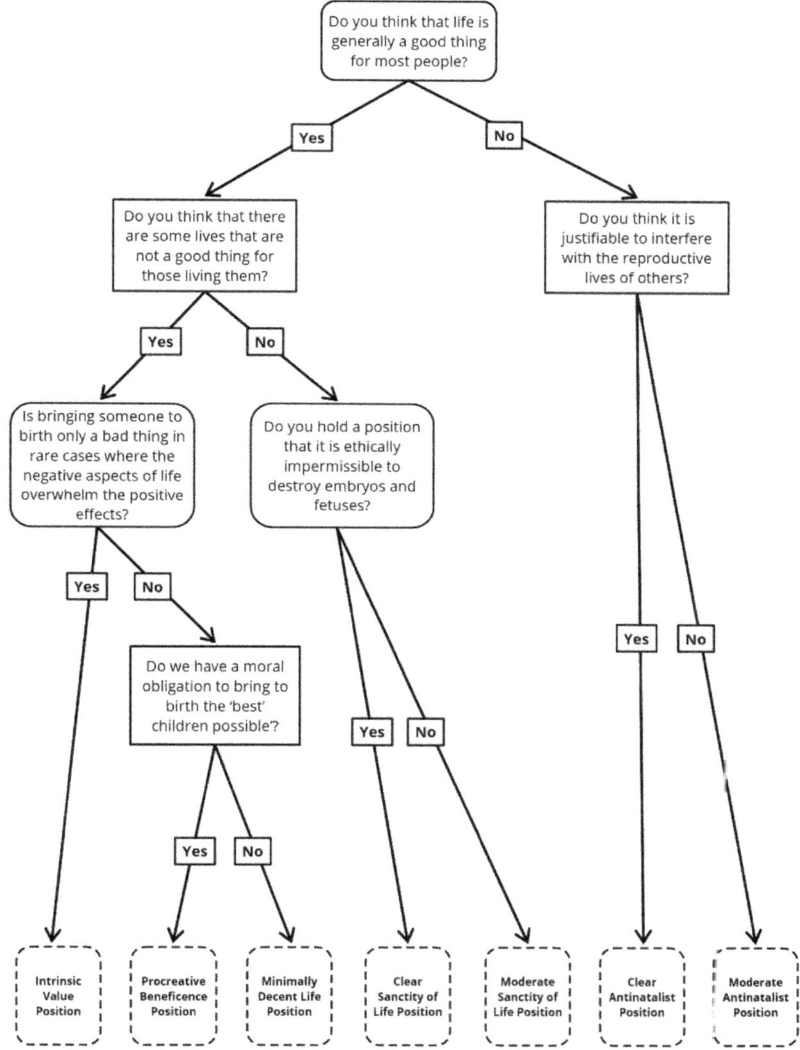


**Figure 3** *Where do you draw the line? Decision tree.*

## Clear Sanctity of Life Position

This position claims that all human lives are highly valuable, including embryos and fetuses. This position assumes that reproduction is a very good thing, perhaps on the basis that it is God's will. Those holding this position are likely to argue against the sanctioning of abortion or voluntary euthanasia.

## *What does this position measure?*

On this view, it seems that the value of human life does not only come from intrinsic value to the person who experiences this life but from other things, for instance, the value may be seen as coming directly from a god or other deity. Those holding this view may have a belief that there are existing souls that are waiting to be born and thus a choice regarding whether to bring these souls to birth might benefit or harm the interests of these waiting souls.

## *How would the Clear Sanctity of Life Position answer the question, 'Is the life that we intend to create one that is of an ethically acceptable quality?'*

This position is likely to answer this question by saying that all human life has value and is above any threshold that determines the sort of lives it is acceptable to create. On this view, it might be argued that we never do anything wrong in allowing an individual to be born. In fact, we might do wrong in preventing an embryo or fetus from continuing to birth.

## Moderate Sanctity of Life Position

This position represents the sort of claims that are very similar to the Clear Sanctity of Life position in that human lives are viewed as highly valuable. However, it is possible that some people who hold this view that all existing human lives are valuable might not extend this to all early embryos or fetuses. There may be those who oppose euthanasia and view all human life as valuable but might accept the use of IVF and even abortion in some circumstances.

## *What does this position measure?*

As with the Clear Sanctity of Life Position, it seems that the high value put on human life doesn't only come from the intrinsic value to the person who experiences this life but from the value that is put on this life by God or some other deity. It may be that, in some circumstances, those holding these views would accept that this value can be overridden, for instance, by a need to treat infertility or to prevent the death of a pregnant person by ending a risky pregnancy.

***How would the Moderate Sanctity of Life Position answer the question, 'Is the life that we intend to create one that is of an ethically acceptable quality?'***

This position is likely to answer this question in a very similar way to the Clear Sanctity of Life Position claiming that all human life has value and is above any threshold that determines the sort of lives it is acceptable to create. However, there may be some instances where the value of the lives of early embryos or fetuses might be overruled by the interests of existing individuals.

## Intrinsic Value Position

This position assumes that life is generally good for those who experience it but with no obligation to reproduce. It accepts that some lives will, in rare cases, fall below the threshold of acceptability, and in these rare cases, we may have reason to prevent these lives coming into existence or assist existing individuals who view their own lives as of unacceptable value to end them.

***What does this position measure?***

The value that is significant here is the intrinsic value that a life might have to the person who experiences it. It is this self-assessment, or our assessment of this self-assessment in cases of future children, that will determine whether a life does or is expected to fall below the threshold of acceptability.

***How would the Intrinsic Value Position answer the question, 'Is the life that we intend to create one that is of an ethically acceptable quality?'***

The Intrinsic Value Position would answer this question by saying that we only do something wrong in allowing an individual to be born if that individual is likely to have a life that is intrinsically harmful to that person – that is, a life dominated by suffering. This will only be the case in rare, extreme cases.

## Minimally Decent Life Position

This position assumes that being alive is generally a good thing for most people, with no obligation to reproduce, but that many lives, including some lives that are of positive value to those who experience them, fall below what is considered to be the minimally acceptable quality of life threshold. This view is one that we explored in Chapter 4 and that argues that if this minimally acceptable threshold is not met then the child, while not harmed by being alive, is wronged by experiencing a life that doesn't meet this standard.[1]

### *What does this position measure?*

There is an assumption here that we have a duty to individuals who will be born to ensure that they are born in a condition or circumstances that we think allows for a minimally decent quality of life. While the value that the individuals themselves put on the quality of life is part of what is being measured here, it cannot be just this intrinsic value alone that is being measured. Those who hold this view argue that while someone may have a life they value, a life that is an overall positive experience to them, it may still be wrong to bring such a life into being if it does not meet a minimum standard of 'things that make human lives good'.[2] The reasons for this must be non-personal, that is reasons not affecting the welfare of the particular child in question, as it has already been argued that the minimally decent threshold is a higher threshold than simply requiring that a life is of intrinsic positive value to an individual. What is being measured here must be more than person-affecting value and must be some kind of impersonal value that, while it does not affect any individual, is somehow important.

### *How would the Minimally Decent Life Position answer the question, 'Is the life that we intend to create one that is of an ethically acceptable quality?'*

For the Minimally Decent Life Position, it is wrong to bring to birth a child who will have a life dominated by negative experience, but it is also wrong to allow a child to be born with a life they themselves are likely to value but that falls short of what is considered to be a reasonable or minimally decent standard. This standard will be higher than the threshold between an intrinsically valuable and intrinsically harmful life.

## Principle of Procreative Beneficence Position

This position assumes that life is generally a good thing for those who live it. For some formulations of this position that postulate the existence and importance of impersonal harm and benefit, this may imply an obligation to reproduce. This Principle of Procreative Beneficence Position argues that where choice is possible, we should always choose to bring to birth the 'best' children possible and to do otherwise is morally wrong.

Those who put forward this view agree with the Intrinsic Value Position that we only harm the individual if we create a life dominated by suffering. However, this position puts the threshold higher than the Intrinsic Value Position in many instances. This is because where it is possible to choose to bring to birth a 'better' life, then the alternative possible 'worse' life will fall below this threshold of acceptability, even if it is a life that is likely to be valued by the person who lives it.

### *What does this position measure?*

Those who hold this view accept that there are no person-affecting or welfare-affecting reasons to justify the position they take. They accept that if a life is created that the person experiencing this life is likely to value, then we have not wronged this person by bringing them to birth. As a result, this position takes other things into consideration apart from the welfare of particular individuals. These other things are impersonal reasons to do with making the world a worse place than it needs to be[3] in terms of cumulative notions of well-being rather than the welfare of individual human beings.

### *How would the Principle of Procreative Beneficence Position answer the question, 'Is the life that we intend to create one that is of an ethically acceptable quality?'*

Accordng to the Principle of Procreative Beneficence Position, while any life that is an overall positive experience for the person experiencing it is an ethically acceptable life in terms of person-affecting concerns, it is still unacceptable to create these lives if a 'better' life could be created instead. This threshold will be higher than the threshold between an intrinsically valuable and intrinsically harmful life.

## Moderate Antinatalist Position

The moderate Antinatalist Position argues that being alive is generally not good or at least has a high risk of being not good, and as a result, we should avoid reproduction where possible. However, on the moderate version of this position, because of respect for the reproductive autonomy of individuals, these views will not be enforced on others. This might mean that those holding these views would try and influence and persuade others that their position is the right one to take but would not support attempts to prevent others from reproducing.

### *What does this position measure?*

There are a number of different positions here. The first is the argument that we all suffer unnecessarily by being brought to birth and that this suffering should be avoided by avoiding reproduction.[4] The second is that while we might think we value our lives and see them as positive overall, we are deluded in doing so, and actually, our lives are not the positive experience we perceive them to be.[5] The final main argument here is that even if some people do value their lives and find them a positive experience, there is a high risk, when reproducing, that the child created will have a life of great suffering.[6] All these arguments focus quite clearly on the welfare of future children, although the second argument assumes that our own self-assessment cannot be trusted.

There are other arguments for Antinatalist positions based on the environmental impact of human reproduction, but these appear to be secondary to the concerns for the welfare of future lives.

### *How would the Moderate Antinatalist Position answer the question, 'Is the life that we intend to create one that is of an ethically acceptable quality?'*

The Moderate Antinatalist Position would answer this question arguing that all human life either does fall below the threshold of what an acceptable quality is or risks falling below this threshold.

## Clear Antinatalist Position

The Clear Antinatalist Position argues that being alive is generally not good or at least has a high risk of being not good, and as a result, we should avoid reproduction where possible. Those holding this more forceful version of this position may argue that the prevention of suffering of new people is so important that as well as remonstrating with others about the undesirability of reproduction, they may wish to attempt to prevent reproduction where possible perhaps by withdrawing funding for assisted reproduction or even stronger interventions.

### *What does this position measure?*

As with the Moderate Antinatal Position, while there are concerns about the environmental impact of human reproduction, the main thing that is being considered in this view is the experience and welfare of future human beings.

### *How would the Clear Antinatalist Position answer the question, 'Is the life that we intend to create one that is of an ethically acceptable quality?'*

The Clear Antinatalist Position would answer this question arguing that all human life either does fall below the threshold of what an acceptable quality is, or risks falling below this threshold.

## The problem of measuring things other than how a future person may experience their life

Clearly, the positions I have outlined above are generalized and lack detail. However, they do indicate the spectrum of positions that are taken on this issue of where we draw the threshold between lives it is seen as ethically acceptable to create and lives that it is claimed we have an ethical duty to prevent. I include this outline of these general positions for two reasons. The first is to encourage you to determine where you stand on this issue and why you take the particular position you do. It may be that you take a slightly hybrid view between positions or a modified version of one of these positions, but by asking yourself the questions in the decision tree and reading these descriptions of the positions, you may be able to add detail to the position you take.

However, the second reason for including these descriptions of these positions is to illuminate an issue that I find of great concern. It is only the Intrinsic Value Position and the two variations of the Antinatal Position that focus *only* on the welfare of the future individual. All other standpoints use things other than the welfare of future lives to determine which side of the threshold of acceptability a life is deemed to fall. Both versions of what I have called the Sanctity of Life Position involve considerations of the interest of a higher power or deity, and thus, holding such a view usually arises from a particular religious or spiritual belief. This will be problematic for those who do not share this particular belief as there is an assumption here that the will of a divine being is more important than the welfare of individual humans. Regulations based on this point of view will be difficult to justify in a pluralistic society particularly if these regulations undermine individuals' choices based on a particular religious belief that is by no means universal.

The Minimally Decent Position and the Principle of Procreative Beneficence Position also take a value to be more important than the individual welfare of future individuals. Both these positions ultimately set the threshold of what is an acceptable quality of life for future lives on factors that are not person-affecting, often around issues of impersonal harm that affect a whole society rather than any individuals. In both these positions, this threshold between what is considered to be an acceptable and unacceptable quality of life will be higher than the threshold between an intrinsically valuable and intrinsically harmful life.

In Chapter 3, I talked about the fact that I am uncomfortable with the terms 'worthwhile' and 'unworthwhile' lives or 'lives not worth living'. My discomfort about using these terms has do with the issue that these terms could very easily be interpreted as implying that the lives of others are *externally* judged as worthless and, thus, that those who have these kinds of lives are not valued in the way that other lives are valued. This is not the way that I and many others have used these terms. I have used these terms to indicate a threshold between lives that are of positive value to those who live them (or are expected to be of positive value to those who live them) and lives that are so dominated by negative experience that they are a harm to that person. To clarify this and to remove this possible ambiguity around these terms, I suggested that in this context, we use the terms intrinsically valuable and intrinsically harmful instead.

There will be those who defend the use of the terms 'worthwhile' and 'unworthwhile' lives or 'lives not worth living' and argue that these terms do not imply any external judgement of the sorts of lives that are in question here.

I agree that the way in which these terms have been used in the bioethical literature has often been synonymous with what I wish to represent with the terms intrinsically valuable and intrinsically harmful. However, as we can see from the setting out of the different positions on the threshold that can be drawn between what is seen as an acceptable and unacceptable quality of life, there are positions that do measure other things than whether the person in question is likely to have a life of positive value.

Both the Principle of Procreative Beneficence and the Minimally Decent Positions agree with the Intrinsic Value Position that a life would need to be overwhelmed by negative experience to be considered *intrinsically* harmful. However, while they accept this threshold between intrinsically valuable and intrinsically harmful lives, this is not where they ultimately set the threshold that should be used to guide reproductive choices. Both positions set this threshold higher than this intrinsic value/harm distinction. On the Principle of Procreative Beneficence Position, the line that is drawn between acceptable and unacceptable reproductive choices is based on considerations of whether the life created is the 'best' one that can be created (where choice is possible). Minimally Decent Positions draw this line slightly differently, but also above this intrinsically valuable/harmful threshold requiring prospective parents to ensure that the lives they create are not just of positive value to those who live them, but that they are lives that exceed this requirement. Of course, what is meant by the 'best' child or what counts as a minimally decent quality of life is something that will be highly subjective and very difficult to agree on.

My own position of the Intrinsic Value Position has its own practical problems with trying to decide when a life is likely to be valuable or harmful to those who live them. The Principle of Procreative Beneficence and Minimally Decent Positions also face practical problems of deciding what counts as the 'best' or 'minimally decent' life and how we might be able to assess this before birth. But it is not just a practical problem of classification that these two positions face. I argue that any position taken here that wishes to impose a different standard rather than the distinction between intrinsically valuable and intrinsically harmful lives is highly problematic. Being concerned that the lives we are creating do not meet a particular standard, not based on the welfare of the individual who will experience that life but based on other concerns, is not only practically problematic, but also fundamentally problematic.

As we saw earlier, Savulescu has rejected the idea that arguing that there is a moral obligation to bring to birth the 'best' child possible is tantamount to eugenics. He argues that eugenics is a *'public interest* justification for interfering with reproduction' 'to promote social goods'.[7] He suggests that the acceptance of a moral obligation to bring to birth the 'best child' possible is a private enterprise and should therefore not be conflated with eugenics. Harris was less reluctant to embrace the term eugenics for his very similar arguments for a moral obligation to avoid disability and to bring to birth the 'best' children possible.[8]

There are different kinds of eugenics. *Authoritative* eugenics involves the state attempting to influence who will be born, whereas *liberal* or *new* eugenics is usually conceptualized as leaving it to prospective parents to make decisions to utilize techniques such as pre-implantation genetic testing – and in the future, genetic editing – to create what they consider to be the 'best' child.[9] Eugenics can also be described as positive or negative. *Positive* eugenics involves persuasion and incentivizing what are considered to be 'good' reproductive choices. This might involve encouraging reproduction in some of the population. *Negative* eugenics involves coercion that aims to reduce reproduction in individuals who are not seen to have desirable traits.[10] Negative eugenic measures might involve discouraging reproduction for some individuals and may even involve forced sterilization or euthanasia to prevent individuals or groups from reproducing.

It is interesting that, more recently, Savulescu has embraced the term eugenics in his work. In a recent interview, a discussion of the creation of 'designer babies', where parents may choose the characteristic of their children either by embryo selection or genetic enhancement, is followed by the interviewer's question, 'Isn't this creepily close to eugenics?'. In response Savulescu says,

It *is* eugenics. But it's a different kind of eugenics, and it is already practiced. It's sometimes called 'liberal eugenics,' when people make decisions about the sorts of children they have. Genetic testing during pregnancy is eugenics. Testing for Down syndrome, cystic fibrosis – those are sorts of eugenics. What was wrong with eugenics in the past was that it was forced onto people. It wasn't for the benefit of the offspring or what parents wanted. It was to bring about a racist, social-Darwinist view of the state.

We can see here that Savulescu is aligning his claim that there is a moral obligation to bring to birth the 'best' child possible with this idea of 'liberal' or 'new' or 'positive eugenics'. While he appears to be embracing the term eugenics, he is also quick to distance the sort of eugenics that he might support from the authoritative, negative eugenics that we think about when we think about the historic Nazi atrocities and forced sterilizations of marginalized groups in society.

We also see the explicit rejection of authoritative, negative eugenics when Savulescu and Harris both emphasize that, while choosing against disability and other characteristics to bring to birth the 'best' child possible is a strong obligation, no one should be compelled to fulfil this proposed moral obligation.[11] Thus, on face value, these arguments are put forward as attempts to prevent harm to future children and prospective parents but without the negative authoritative sting of the 'bad' kind of eugenics.

It may well be that Harris, Savulescu and others view their arguments as very different from the historical monstrosity that the word 'eugenics' often conjures. However, on closer inspection it is extremely difficult to distance both the Procreative Beneficence Position and the Minimally Decent Position from these difficult associations.

Any argument that condemns reproductive choices based on anything other than the intrinsic welfare of future individuals will be left wide open to accusations of eugenics of the worst kind. Ideas of what is the 'best' human life or what is a minimally decent quality of human life that are based on something else than whether a life is likely to be of positive value to the person who lives it are always going to be difficult to justify and difficult to distance from the idea of authoritative, negative eugenics.

Proponents of these views may temper these positions by insisting that respect for reproductive autonomy means that any moral obligations to choose to bring to birth the 'best' child possible or a child whose welfare is considered to be a minimum standard of welfare above what might be considered to be an intrinsically valuable life should not be enforced. However, putting forward these arguments in a society that has established regulations that restrict and attempt to influence reproductive choices in ways that often align with these arguments, effectively lends support to these regulations. By providing arguments that appear to provide theoretical, high-level academic support for these regulations, those who argue for a moral obligation to avoid bringing to birth children with disabilities are supporting regulations that, I argue, can be viewed as authoritative, negative eugenics.

I have argued that deciding what is an acceptable level of welfare based on anything other than the value that the person experiencing this life is likely to put on this life is highly problematic. As we have seen, there are those who argue that we have reasons above and beyond the welfare of individuals to make particular reproductive choices, to ensure that lives that are created are the 'best' they can be or to meet a minimum expectation of welfare, and there is a danger that lives might be assessed as 'worthwhile' or 'unworthwhile' not only based on the expected welfare of these lives, but also on these other impersonal values. My hope is that using the terms 'intrinsically valuable' and 'intrinsically harmful' will remove this ambiguity.

Further, these terms are not only used to express this notion of intrinsic value, but may also be used to express an external judgement of the value of a particular life.


## Conclusion

When we ask ourselves the question 'Is the life that we intend to create one that is of an acceptable quality?', we imagine a threshold between what we consider to be an acceptable and unacceptable quality of life. Exploring where these thresholds may be drawn, depending on our answers to a number of pertinent questions, not only provides an insight into our own positions on this issue, but also illuminates the motivations of these different approaches. Illuminating these motivations is really helpful in our evaluation of arguments that focus on the welfare of future children. For example, by exploring these thresholds in some detail, it becomes clear that the Procreative Beneficence Position and the Minimally Decent Position draw this distinction between what is considered to be an acceptable and unacceptable quality of life based not just on whether a life is likely to be a positive experience for the person who lives it, but on external judgements of what is an acceptable quality of life. External judgements of what is considered to be an acceptable quality of other's lives will always be highly problematic. As a result, I have argued the positions that make these kinds of external judgements are very difficult to justify generally and, particularly, if we maintain our focus on the welfare of individuals.

# CHAPTER 7
# WHERE DOES CURRENT REGULATION ON THE WELFARE OF FUTURE CHILDREN DRAW THE LINE?

Concern for the welfare of future children has widespread support and is required by regulation around the world. As we have seen, when we make decisions based on the welfare of future children, we do so by attempting to imagine thresholds between the sort of lives we feel comfortable in creating and the creation of the sort of lives that makes us feel uncomfortable. While these decisions are often explicitly taken based on our assessment of the expected quality of life of the individual who experiences this life, other factors such as our religious beliefs, concerns regarding burdens on society and negative biases and social norms around disability contribute to and, in many cases, dictate where we draw this line. As a result, where this threshold lies and how it is arrived at is usually not explicit or easy to pin down. In this chapter, I will explore the way that current regulation in this area encourages this threshold setting and argue that this current regulation is unhelpful to those making these decisions, is non-transparent and thus difficult to challenge or discuss, reinforces unhelpful and stigmatizing attitudes, undermines individual reproductive choices unjustifiably and is ineffective in its goal of protecting the welfare of future lives.

**Current regulation threshold: Question 1**

Question 1: Are we justified in attempting to evaluate the potential parenting ability of those trying to access fertility treatment (e.g., disabled people or individuals with past criminal convictions) and prevent access in some cases?

As we saw in Chapter 1, in the UK, the Human Fertilisation and Embryology Act 1990 (HFE Act 1990) requires all licensed centres providing fertility treatment to consider the welfare of any resulting child when providing any services to assisted conception and pregnancy. The original wording of this requirement in the act in 1990 was as follows:

> A woman shall not be provided with [fertility] treatment services unless account has been taken of the welfare of any child who may be born as a result of the treatment (including the need of that child for a father), and of any other child who may be affected by the birth.[1]

At this time, the Human Fertilisation and Embryology Authority (HFEA), the group set up to oversee the regulation of fertility treatment in line with the HFE Act 1990, interpreted this requirement to mean that centres should withhold treatment where there is 'reason to suppose there is a risk of harm to the child',[2] which, without further guidance, would seem to encourage the setting of the threshold for treatment very high. There was no further explanation of what was meant by harm or how serious this harm would need to be to be relevant here. As a result, 'red flags' might be raised by any information about physical or mental health, previous convictions, drug or alcohol use, behaviour or information on living conditions that were observed or volunteered by prospective parents or their general practitioners (who were routinely contacted to comment on this issue of the welfare of the child). It was left to clinics to decide how reliable this information was and how much weight to give it when deciding who should have access to treatment. With very little further guidance provided to clinics, it seemed inevitable that this guidance would be interpreted in different ways by different clinics.

The requirement to consider the 'need of that child for a father' meant that many National Health Service (NHS) providers of licensed fertility treatment routinely restricted treatment to relatively young 'healthy' heterosexual couples.[3] This reflected the social norms of the time and concern about increasing numbers of single-parent families. The last-minute adding of the clause of the need for the child of a father was done in response to these concerns that deliberately creating children without fathers was not something that the government wanted to encourage. These restrictions were made from concern for the welfare of any children born as was evident in this comment from the then Lord Chancellor, Lord Mackay, when he said,

> through counselling and discussion with those responsible for licensed treatment, (single women) may be discouraged from having children

once they have fully considered the implications of the environment into which their children would be born or its future welfare.[4]

In 2004, the HFEA launched a review of its Welfare of the Child guidance and in 2005 ran a public consultation,[5] seeking views for revising this guidance. The HFEA noted that

> A strong message that came across during the consultation was a desire for clearer guidance on how clinics should interpret the welfare of the child provision. Clinics want to know exactly what steps they should take to meet their legal responsibilities. Patients also want clear guidance, to enable them to understand the criteria against which they are being assessed.[6]

There was concern that the guidance put forward by the HFE Act was too vague to be helpful in these assessments. Around this time, it was claimed that 'recent research has shown the welfare of the child to be a slippery concept' which 'may be ineffective and permit less legitimate and discriminatory activities, such as the exclusion of certain social groups'.[7] There was also evidence internationally that where similar requirements to consider the welfare of the child for fertility treatment are required 'screening practices and determining access to ART [Artificial Reproductive Technologies] showed substantial variation in assessment practices and a high risk of inappropriate denial of treatment'.[8]

As a result of the 2005 review and public consultation, the UK Welfare of the Child provision has evolved with amendments to the HFEA Code of Practice in 2007[9] and formal amendments to the HFE Act in 2008. These changes effectively lowered the threshold for treatment based on the Welfare of the Child provision.

This amended Code of Practice emphasizes that rather than for looking for reasons not to treat patients, there is a presumption in favour of providing treatment[10] unless there is a 'risk of serious harm or neglect'.[11] There is no longer a requirement to routinely contact the general practitioner of service users, and more detail on what is considered to be a risk in terms of this serious harm or neglect is given as follows:

> The centre should consider factors that are likely to cause a risk of significant harm or neglect to any child who may be born or to any

existing child of the family. These factors include any aspects of the patient's or (if they have one) their partner's:

(a) past or current circumstances that may lead to any child mentioned above experiencing serious physical or psychological harm or neglect, for example:
   (i) previous convictions relating to harming children
   (ii) child protection measures taken regarding existing children, or
   (iii) violence or serious discord in the family environment

(b) past or current circumstances that are likely to lead to an inability to care throughout childhood for any child who may be born, or that are already seriously impairing the care of any existing child of the family, for example:
   (i) mental or physical conditions
   (ii) drug or alcohol abuse
   (iii) medical history, where the medical history indicates that any child who may be born is likely to suffer from a serious medical condition, or
   (iv) circumstances that the centre considers likely to cause serious harm to any child mentioned above.[12]

The amendments to the HFE Act in 2008 removed the formal requirement to consider the 'need of that child for a father', substituting this with 'the need of that child for supportive parenting'. As a result, UK clinics no longer restrict access to treatment to heterosexual couples.[13] It is important to note that while inclusion of the clause for the 'need of that child for a father' in the wording of the HFE Act and the HFEA Code of Practice, was motivated by concerns for the welfare of the child, this concern is not supported by evidence that children born to single women and lesbian parents fared any worse than those born to a heterosexual couple; in fact, there is now compelling evidence that this is not the case.[14]

Aside from this 'need for a father' clause, if we think about the UK Welfare of the Child provision in terms of thresholds between what are predicted to be acceptable and unacceptable levels of welfare, it does appear that this provision has moved from encouraging a relatively high threshold, where *any* harm at all might be considered a reason to question or even refuse treatment, to a much lower threshold where there is a presumption that treatment will be provided unless there is evidence of what is considered to be a '*serious*' or '*significant*' harm.

## Current regulation threshold: Question 2

Question 2: Should we allow prospective parents using IVF to implant an embryo with a condition considered a disability? For example, should a deaf person be allowed to implant a 'deaf' embryo?

As outlined in Chapter 1, in the UK the Human Fertilisation Act 2008 explicitly prohibits the implantation of embryos that will result in the creation of children with 'a serious physical or mental disability' 'a serious illness' or 'any other serious condition' if there are other embryos that are unaffected by these conditions.[15]

Pre-implantation genetic testing screens embryos for specific genetic 'defects' involving a single gene. In the UK, currently, IVF embryos can be screened for 1442 conditions.[16] Individuals and couples are given access to pre-implantation genetic testing in the UK when there is 'a particular risk that the embryo to be tested may have a genetic, mitochondrial or chromosomal abnormality, and the Authority is satisfied that a person with the abnormality will have or develop a serious disability, illness or medical condition'.[17] Whether there is a risk of these conditions will be assessed on the gamete donor's history.[18]

This regulation prohibits the use of a gamete donor or the implantation of an embryo where there is 'significant risk' that this choice will result in the creation of a child with a 'serious physical or mental disability', illness or conditions. However, as this clause says that these embryos or donors 'must not be preferred' where there are no 'unaffected' embryos, these affected embryos 'may be transferred'[19] 'subject to consideration of the welfare of any resulting child'.[20]

The implication here is that lives with serious physical or mental disability are undesirable lives, at least in contrast with 'unaffected' lives, and thus, the threshold of acceptability is put at this level of 'significant risk' of 'serious' disability. As a result, the UK's current position on this issue seems very much in line with the current position on access to fertility treatment where the language is that of 'significant risk' and 'serious harm'.

## Do these regulations do the job they set out to do, that is, do they enable decision-making that safeguards the welfare of future children?

In the UK, the HFE Act 2008 regulates both access to fertility treatment and use of pre-implantation genetic testing with regard for the welfare of future children using a threshold of risk of 'serious' or 'significant' harm or disability. As we saw in Chapter 1, this approach is also reflected strongly in international regulations. As a result, those making decisions in these two areas are encouraged to answer the question 'Is the life that we intend to create one that is of an ethically acceptable quality?' by attempting to assess whether the life created will be one that will be subject to risk of 'serious' or 'significant' harm or disability. The implication is that any life assessed to be subject to this level of risk would be below the threshold of acceptability.

At first glance, these regulations seem to meet the protective need that many of us feel when trying to ensure that children born with the assistance of others meet a 'minimum' standard of what is acceptable. We feel a responsibility to ensure that we are not instrumental in enabling the birth of children whose quality of life is thought to be problematic. In an attempt to provide more guidance to those making these decisions and to avoid prevention of a birth for what might be seen as minor reasons, the requirement here is that the harm that is likely to be caused is 'serious' or 'significant'.

However, identifying what is meant by 'serious' or 'significant' harm or disability is hugely challenging. Depending on the life experience, biases and influences of those making these decisions, the category of serious harm or disability could cover a huge number of conditions and circumstances. One person's serious harm will be another's minor harm, and this might change depending on the condition or the circumstances being considered. There is a risk that where the threshold is placed will change dramatically from one case to another and one clinic to another.

Perhaps due to these difficulties in agreeing on and articulating what counts as serious harm or disability, the HFE Act 2008 and the HFEA Code of Practice gives little indication of what is meant by these terms or even what is meant by the term 'disability'. As we have seen, the HFEA Code of Practice that provides guidance for access to fertility treatment based on the Welfare of the Child provision does provide a list of the sort of factors that should be considered when making this assessment of potential 'serious' harm. This list includes a history of 'convictions relating to harming children', 'child protection measure's, 'violence', 'mental or physical conditions', 'drug and alcohol abuse' and the risk of the resulting child having a 'serious medical condition.'[21] While this list is useful in providing an idea of the sort of things that might be considered here, the huge scope of these factors for consideration, unfortunately, does not do a great deal to reduce the amount of subjectivity that is involved in making these decisions. For instance, while, of course, convictions, child protection measures and evidence of violence or drug or alcohol abuse will be relevant here, there is no indication of how much weight to give this information and how much weight later rehabilitation should be given here. We know that those with a history of violence, child neglect and substance abuse are often victims themselves of childhood violence, abuse and neglect.[22] This behaviour can be a response to this trauma, and with help and support, individuals can overcome these issues. Similarly, assessing which 'mental or physical conditions' are 'likely to lead to an inability to parent effectively throughout childhood for any child who may be born'[23] or whether 'medical history indicates that any child who may be born is likely to suffer from a serious medical condition'[24] is highly subjective, particularly without any further detail or examples.

Further, in the case of regulation of pre-implantation genetic testing, the HFEA Code of Practice suggests that providers of pre-implantation genetic testing should consider the following factors when deciding whether this procedure is appropriate, that is whether the condition that individuals seek to screen for is something that is a risk regarding 'serious disability, illness or medical condition':

(a) the views of the people seeking treatment in relation to the condition to be avoided, including their previous reproductive experience

(b) the likely degree of suffering associated with the condition

(c) the availability of effective therapy, now and in the future

(d) the speed of degeneration in progressive disorders

(e) the extent of any intellectual impairment

(f) the social support available, and

(g) the family circumstances of the people seeking treatment.[25]

This gives some more detail regarding what sort of factors might be considered when assessing whether a condition constitutes a 'serious disability, illness or medical condition'[26] in the context not only of allowing access to pre-implantation genetic testing, but also deciding which embryos should not be preferred for implantation. However, again, these are very general descriptions, with no indication of how we assess the seriousness of a condition and how this assessment should take account of other possible factors that might change this assessment. For instance, for many of us, 'serious disability' would not include deafness, given we know that those with this condition are likely to value their lives as much as anyone else, and for many, being part of Deaf culture[27] can be an enriching part of an individual's quality of life. However, we know that, with regard to this regulation, deafness is seen as a serious disability as it was concerns about using donor gametes and pre-implantation genetic testing to select for deafness that prompted the addition of the clause preventing the implanting of affected embryos.[28]

The HFEA has licensed pre-implantation genetic testing for a number of cancer susceptibly genes, including BRCA1 and BRCA2. Inheriting the BRCA1 and BRCA2 gene raises the risk of breast cancer in women from around 13 per cent in the general population to around 45 per cent–72 per cent and ovarian cancer from about 1.2 per cent in the general population to about 11 per cent–44 per cent.[29] Given that breast and ovarian cancer are both adult-onset conditions and the requirement that pre-implantation genetic testing only be used to screen for a risk of 'serious disability, illness or medical condition',[30] this suggests that the HFEA definition of serious disability includes a predisposition to some adult-onset conditions. Remembering that these are predispositions and not definite conditions and recognizing that the risk for any one individual depends on a number of factors, this would seem to set the definition of serious disability very wide. While it may be understandable why many individuals and couples wish to test embryos for adult-onset predispositions to cancer where possible, this does seem to indicate that definitions of 'serious disability, illness or medical condition'[31] used as part of the regulatory framework here are very elastic.

As a result, while regulation in these areas appears, at first glance, to be responsible and supportive, without further guidance, this regulation ultimately appears to leave the healthcare professionals involved to make their own judgement about what is a serious risk of harm to the welfare of resulting children. The high level of subjectivity that is involved is likely to lead to inconsistent decisions between clinics and even between patients, and it may not just be the 'facts' that influence these decisions, biases about certain conditions or cultural or social factors are likely to influence where this threshold is placed between acceptable risk and unacceptable risk of harm.

As we discussed earlier, we are all biased and even being aware of this fact and trying to mitigate this bias is usually ineffective. As a result, even the extremely well intentioned are biased.[32] In other areas of life, as we saw earlier in this book, we often attempt to lessen the effects of this inevitable bias by employing ways of controlling the subjectivity of decisions made. However, in the current Welfare of the Child provision, as it is applied to both access to fertility treatment and use of pre-implantation genetic testing, there are very few limits on the scope of subjectivity and thus there is an inevitably high risk of the influence of bias in decision-making.

While the role of bias is likely to affect all decision-making in this area, it is perhaps most evident when it comes to single women's access to fertility treatment even after the removal of the 'need of that child for a father' clause in the Welfare of the Child provision in the HFE Act 2008. In 2014, a study observed the Welfare of the Child provision in practice found that 'single women still appeared to be regarded by some as potentially problematic patients and parents, less because of the child's perceived need for a father, and more on the grounds of their motivations and ability to cope with parenthood'.[33] Despite there being strong evidence that children's social and emotional development is not negatively affected by the absence of a father,[34] existing attitudes towards single women parenting seem to engender feelings of concern when it comes to the welfare of future children.

When it comes to access to fertility treatment (at least in the UK), this risk of inconsistent subjective decisions is exacerbated by the fact that only those who are identified as raising 'concerns' (either through volunteering information or issues observed by clinic staff) are likely to be scrutinized in terms of the Welfare of Child provision. Clinics no longer routinely contact patients' general practitioners to ask about any concerns regarding the welfare of the child.[35] This means that it is the honest patients who divulge past convictions, mental and physical health diagnoses and so on who are more likely to be scrutinized on this basis. Those who choose not to divulge this information are much less likely to be investigated. Similarly, patients who present with conditions that are obvious, such as a certain physical or mental health conditions, may find that they are asked more questions about their ability to parent, in line with the current Code of Practice, than others who may also have physical or mental health challenges that are less obvious. For instance, a woman with a diagnosis of a learning disability who brings her mother to support her is likely to have more extensive scrutiny regarding the welfare of her child than a high-functioning, middle-class alcoholic who is able to mask their condition when attending clinic. Who you are, your social status, how you present yourself and what you divulge will inevitably have an impact on how much you are investigated under this Welfare of the Child provision.

There is evidence that it is usually a group of people who make these decisions about access to fertility treatment based on the Welfare of the Child provision. Rather than being one consultant taking this decision, the decisions are usually made in a clinical team meeting by consensus or by an ethics committee.[36] Those making these decisions often recognize the difficulties of subjectivity here and feel that a group decision would be less likely to be influenced by prejudice and bias. One clinic doctor expressed this when he said,

> because we all have our prejudices, even though we like to think we're all good honest people, but there are things that I approve of and don't approve of, and only by taking into account a lot of people's feelings do we get it right.[37]

However, while using a group to help make these decisions may mitigate some of the subjectivity issues here, it is unlikely to remove them and the influence of bias and prejudice significantly. There are a number of reasons for this. Given the hierarchical structures of healthcare, it is likely that some opinions will have more influence than others. It has been suggested that in striving for a consensus on these cases, it may be that this could lead to a tendency to conservatism,[38] where those who do not feel it is acceptable to treat in certain circumstances effectively veto access for these cases. This tendency to conservatism may also be exacerbated by the widespread unconscious negative attitudes that many of us have towards disability and the effect this has on our ability to accurately assess the likely quality of life of disabled people or in circumstances that we find challenging. Ultimately, given that our biases are often generally shared ones, particularly around issues such as disability, child protection and criminality, it is likely that a group of individuals will reflect these biases and prejudices in the same way as individuals unless there are other ways to check the influence of these biases.

### The grey area of subjectivity

As a result of this ambiguity in current UK regulation, and others that have a similar reliance on avoiding 'serious' or 'significant' harm or disability, there is a massive grey area of subjectivity when it comes to decision-making on these issues where any, all or even no factors could be considered to meet these criteria of 'serious' harm or disability. If we think about the general potential thresholds that we identified earlier, this lack of clarity and thus flexibility around what is meant by 'serious' or 'significant' disability or harm means that those taking these decisions could take any of these positions on possible thresholds and still work within these regulations (see figure 4).

Those taking the position I take, what I call the Intrinsic Value Position, would argue that for harm to be 'serious' or 'significant' enough to justify infringing on the reproductive choices of individuals, this harm would have to be such that it would be likely to lead to an intrinsically harmful life, one that is dominated by suffering. However, those taking a Minimally Decent Life Position, or a Principle of Procreative Beneficence Position (see Chapter 6 for details of these positions) would argue that the threshold for 'serious' or 'significant' harm for future individuals is much higher. On this view, it may

| Maximum Welfare | *Clear Sanctity of Life Position* |
| --- | --- |
| | *Moderate Sanctity of Life Position* |
| | *Intrinsic Value Position* |
| | *Minimally Decent Life Position* |
| | *Principle of Procreative Beneficence Position* |
| | *Moderate Antinatalist Position* |
| Minimum Welfare | *Clear Antinatalist Position* |

**Figure 4** *The grey area of subjectivity.*

be argued that we have a responsibility to not only prevent terrible lives of intrinsic harm, but also lives that either do not meet a minimum standard or are not the 'best' that we can achieve. While those with an Antinatalist stance may perhaps be unlikely to be involved with fertility services and thus be tasked to make these decisions, if they were called on for their assessment of what is 'serious' or 'significant' harm or disability, it is likely they would suggest a very low threshold where they feel there is good reason to prevent the birth of many or most lives. Individuals taking a Sanctity of Life Position who do not object to fertility treatment or pre-implantation genetic testing on the grounds of destruction of embryos are likely to have a very high threshold of what they consider to be 'serious' or 'significant' harm or disability. Even with the 2008 HFE Act's amended approach that emphasizes an assumption to treat, this highly ambiguous limitation that treatment can be refused in cases of 'serious' harm or disability means that this guidance allows *all* of these different views here to flourish without checks.

Against this, it might be argued that while there may be many different views about what lives are likely to involve 'serious' or 'significant' harm or disability, there will be some agreement as to what sort of cases fall into the two extreme categories here: cases of clear serious harm or disability and cases where there is no harm or disability. On this basis, it might be argued that while there is a grey area in the middle where decisions are subjective and likely to be inconsistent, this area is small.

However, while this idea may seem intuitively attractive, when we actually try and determine what sort of lives would be in these two extreme categories, we can see that this is not the straightforward task we may have assumed it to be.

When we consider examples of lives that clearly do not or are not likely to be subject to serious harm or disability, how do we determine this? Would these be individuals that had no kind of mental or physical condition that can be seen as disabling? Even pinning down what is meant here by a disabling condition is hugely problematic. For instance, is having autism or having a genetic predisposition to obesity or depression likely to cause 'serious' or 'significant' harm or disability? If this seems like an unfair question, then we could consider deafness as this is definitely included in the definition of 'serious' disability when it comes to pre-implantation genetic testing, at least in the UK, as the clause we are focusing on was introduced to avoid anyone using pre-implantation genetic testing to choose to bring to birth a deaf child.[39] While deafness may be something that negatively affects the quality of someone's life (although this may not always be the case) it is unlikely

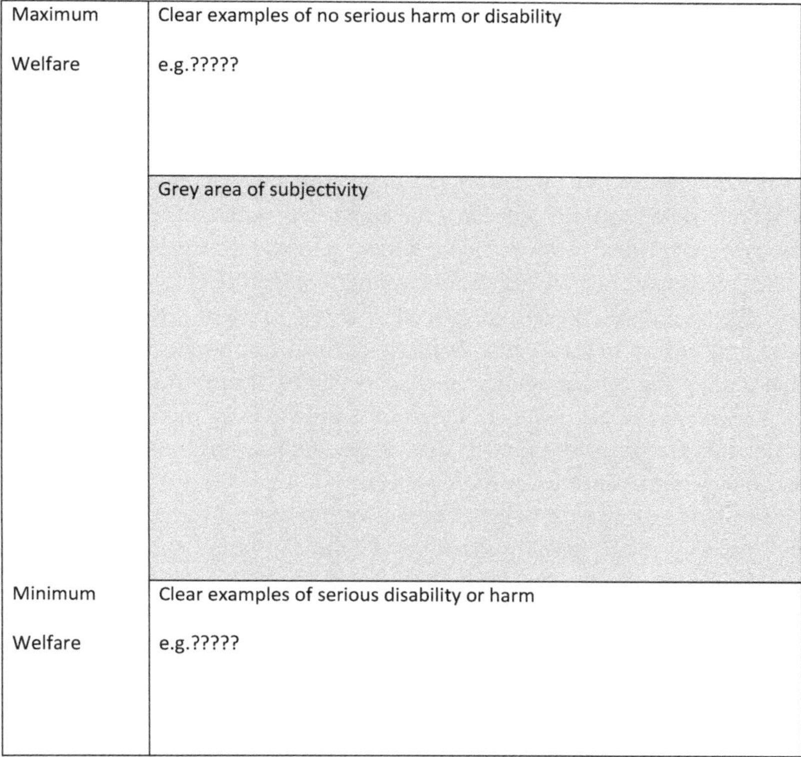



**Figure 5** *Reducing the grey area of subjectivity – what counts as no harm or very serious harm?*

to affect it so dramatically as to undermine the positive elements of life. It seems that similar things could be said about being on the autistic spectrum and having a predisposition for obesity and depression.

If we think about this question in an even wider sense, is it possible to distinguish these kinds of conditions from other characteristics that may make a life more challenging – for instance, being female, gay, transgender, intersex, with a particular racial background, being less attractive than others, being born into poverty? When does a disadvantage become a disability and vice versa? In response, it might be argued that when it comes to prohibiting the preferring of embryos for implantation after pre-implantation genetic testing, we should only prohibit those with 'medical' restrictions on their functioning. It might be argued that other factors such as being gay in a homophobic world or a person of colour in a racist world are factors that could be solved by changes in society. However, it could, of course, be argued that being deaf or living with a condition like achondroplasia, resulting in much smaller than average height, might also be said to be conditions where disadvantages could be addressed through similar social changes. Further, if we attempted to limit our concerns about the welfare of future children to risks of 'medical' notions of disability, then the routine welfare of the child assessment of individuals attempting to access fertility treatment on social grounds would seem misplaced. If trying to avoid creating children who will have a lower quality of life is deemed important, then it seems that attempting to distinguish 'medical disability' from other factors that we know may impact quality of life to similar degrees as 'medical disability' is not only fraught with practical difficulties, but it seems to undermine the legitimacy of this endeavour more generally.

Similarly, trying to provide examples of lives that are clearly below the threshold of serious harm or disability is a difficult task. Emily Jackson, for instance, argues that

> Some genetic conditions are self-evidently serious. If a child is born with Tay-Sachs disease, her nervous system will start to degenerate during her first year of life, and she will die within three or four years. The seriousness of other conditions may be less clear-cut. The [HFEA] Code of Practice acknowledges that whether or not a condition is serious is not only an objective medical fact, but may also depend upon the family's circumstances and their subjective views of the condition.[40]

While we might support individuals' use of pre-implantation genetic testing to avoid conditions in their children that they see as serious and that they would prefer their future children not to have, this commitment to reproductive choice in this situation does not necessarily commit us to reducing the reproductive choices of parents who may take a different view of a given condition, given the subjectivity that may be involved in this assessment. There may be some extreme conditions such as Tay-Sachs that we can come to a general agreement about, but it is likely that these will be a very small subsection of the conditions that are currently identified as 'serious' and that currently, at least in the UK, prohibit the implantation of an affected embryo.

As a result, guidance that focuses on a threshold of 'serious' harm or disability without providing more detailed information about what conditions or challenges might fall below this threshold and what might not doesn't seem to provide much guidance at all when it comes to this challenging decision-making. This is not only problematic for those who have to make these decisions, but it is also problematic for other reasons which we will examine now.

## Safeguarding the welfare of future children

While these regulations may leave the door open to subjectivity and even bias when it comes to decision-making in these two areas, it might well be that this is not a fundamental concern if the regulation and the guidance based on it does the job that it sets out to do – that is, safeguards the welfare of future children.

But there are a number or reasons why we might question the effectiveness of these approaches.

### *Access to fertility treatment*

It seems that the number of individuals who are refused treatment based on this Welfare of the Child provision has remained consistently relatively low.[41] The current regulations, therefore, do little to change who will be born and thus are unlikely to have a significant impact on the welfare of future children. Further, most individuals who attempt to access fertility treatment do so without a great deal of scrutiny around the Welfare of the Child provision, but at least some of these individuals will go on to have children who may face the sort of challenges that would have provoked concern. This might either be because the clinics were not aware of any issues that may have prompted this scrutiny if they were known, or it may just be that conditions arose that risked impacting on the quality of life of this child. Regulations around the welfare of the child will not prevent the creation of children in challenging circumstances as these circumstances often cannot be foreseen.

### *Regulation around the use of pre-implantation genetic testing*

While conditions such as deafness may well be characteristics that many potential parents classify as serious risks and would want to avoid in their children, we know that others, particularly those whose own experience of living in the deaf community has been a positive one, would not classify it as a serious risk. Even those who wish to avoid deafness are unlikely to think that this condition would mean that the life of a deaf child is one that is overwhelmed by negative experiences. However, they may have legitimate reasons why they want to avoid this condition in their own children. I have suggested that individuals are very likely to have a preference to avoid having a disabled child, perhaps due to the impact they believe this condition may have on their own or their families' lives more generally or perhaps due to the widespread social norms that mean that those who have not had personal experience of deafness view this condition in, what some might argue, a disproportionately negative way. However, I have argued that this is a preference rather than a moral obligation or duty to avoid creating a disabled child. It is difficult to justify prohibiting the implanting of a 'deaf' embryo on the grounds of harm avoidance, as, because we have seen, this child is as likely as anyone else to value their life. Thus, choosing not to implant embryos who will be deaf or have another condition that may create challenges, but is unlikely to make a life intrinsically harmful, does not seem to safeguard the lives of any future children. Whether we choose the 'deaf' embryo or the hearing embryo, a child will be born who will be as likely as anyone else to value their life.

## *Why is it ok to implant an 'affected' embryo if there is no other choice?*

It is important to note that current UK regulation around pre-implantation genetic testing says that embryos likely to develop a disability 'must not be preferred to those that are not known to have such an abnormality',[42] and thus, choosing an embryo that is likely to develop a disability is only prohibited when there are 'unaffected' embryos that could be implanted in their place. This sentiment, of course, speaks to the intuition that is behind arguments around the Principle of Procreative Beneficence, that where choice is possible, we should choose to bring to birth the 'best' child possible. However, while this chimes with many of our intuitions around disability, we have seen how difficult it is to provide strong justifications for this position.

I have argued that no convincing account has been put forward that identifies the harm done by a choice to bring to birth a child with a disabling condition such as deafness. We know that those living with deafness and other disabilities are as likely as anyone to value their lives and, thus, insisting otherwise is to dismiss the lived experience of those living with these conditions. There is also evidence that social norms and unconscious responses to disability mean that non-disabled people often have a negatively skewed idea of what living with a disability is like.

However, even if we put these difficulties aside and assume for the moment that a convincing account can be given of the harm that it is suggested is done by choosing to implant a 'disabled' embryo, the way that this law is formulated is highly problematic. The UK clause we have been considering stipulates that you cannot implant an embryo likely to develop a serious disability – presumably, as this is either bad for the person or bad for society in some way (both of these claims being very difficult to uphold). However, if there are no unaffected embryos, there is no prohibition on bringing to birth embryos who are likely to develop what are considered to be serious disabilities 'subject to consideration of the welfare of any resulting child'.[43] If all embryos created are 'affected', then our concerns about bringing a child with a serious disability to birth seem to change. No longer is being 'affected' with what was considered to be a condition that would cause 'a serious physical or mental disability' 'a serious illness' or 'any other serious condition'[44] a reason to prohibit implantation, and embryos with such conditions are then assessed under the Welfare of the Child provision. It seems that here the law sets the threshold between what is considered to be an acceptable and unacceptable quality of life at serious disability or illness but then throws out this threshold if there are no embryos above the threshold to choose from, falling back on the equally vague Welfare of the Child provision in these cases. This appears to take an already highly subjective and problematic legal requirement and then undermine it further.

## Current regulation threshold: Question 3

Question 3: Is routine screening for Down syndrome in pregnancy ethically acceptable even if there is evidence that individuals may feel pressure to accept this screening?

Routine prenatal screening differs from our first two questions in a number of ways. With prenatal screening, a pregnancy has already been established. As a result, if a pregnant person decides they do not wish to bring to birth a child with Down syndrome, this will involve a termination of pregnancy which may have a negative psychological impact on that pregnant individual.[45] As we saw in Chapter 1, it will often be argued that the motivation for routine prenatal screening of this kind is not to prevent the birth of children with Down syndrome in an attempt to prevent harm to that child and society in terms of additional financial and other 'burdens', but to empower pregnant people's choices. As such, routine prenatal screening for Down syndrome does not attempt to set thresholds in the explicit way that regulation on access to fertility treatment and pre-implantation genetic testing does.

### *Empowering pregnant people or harm prevention?*

But, as I argued in Chapter 1, making a case that the routinization of screening for Down syndrome can be justified on the basis that it empowers pregnant people's choices is a very difficult case to make. The routine nature of this kind of screening and the recommendation this seems to imply may mean that routine screening is incompatible with gaining voluntary informed consent.[46] Further, it may be that providing access to prenatal testing for Down syndrome using the same standards of testing that are usually required for genetic testing – that is, that information is balanced and that no pressure is put on someone to accept the offered test – is likely to empower pregnant people's choices much more transparently and directly than routine screening.

Attempting to justify routine prenatal screening for Down syndrome based on public health goals is equally challenging. Targeting those with Down syndrome as a 'burden' on society is not only inaccurate in many cases, but it appears to unfairly single out Down syndrome and seems open to clear accusations of coercive negative authoritative eugenics. As we have already discussed, negative authoritative eugenics is the manipulation of reproductive choices by the state with a goal of furthering what might be called 'social goods'[47] and may include attempts to prevent the birth of individuals who are viewed as having 'undesirable' traits. We know that trying to reduce the number of individuals born with what was seen as the undesirable trait of having Down syndrome was what motivated the early days of screening for this condition.[48] We also know that routinization of screening for Down Syndrome entails a level of coercion that might not be accepted in other areas of healthcare, with consistent evidence indicating that a significant number of pregnant people do not know why they are being screened[49] or are unaware that screening is voluntary.[50] Justifying regulation that appears to be aligned with coercive state-sanctioned eugenics with no evidence of harm to the welfare of individuals will be a significant challenge.

If routine prenatal screening cannot be justified by empowering pregnant people's choices or by public health goals that imply eugenic ideas about those with Down syndrome somehow being undesirable and a burden, then the only possible justification for routinization of screening for Down syndrome is the idea that preventing the birth of individuals with Down syndrome prevents suffering.

## *Harm prevention?*

It might be argued that routine prenatal screening is another example, alongside other screening programmes for breast and bowel cancer, routine vaccination drives or compulsory treatment for serious communicable diseases, where some pressure to accept a screening test or other treatment may be justified to prevent harm, in this case a harm in the form of a future child who it may be thought will be harmed by being brought to birth with a particular disability.

But, as we have seen, establishing that bringing to birth a child with Down syndrome harms anyone is very difficult to do. For those of us who see being alive as a positive thing, this positive value does not seem to be overwhelmed in those with Down syndrome any more than it does in anyone else who may not live with this condition. Although we might have an intuition that having Down syndrome is harmful to the person who lives with this condition, no real sense can be made of this feeling, and although we might be tempted to ask whether someone would be better off without Down syndrome, this is not a possibility when it comes to this particular individual. Unless we assess those living with Down syndrome to be likely to have a life completely dominated by suffering, which is not borne out by the lived experience of those with this condition, we cannot justify this different approach to consent to screening using harm prevention.

If we want to justify routine screening for Down syndrome based on preventing harm to the resulting child, then the threshold we draw between what is considered to be acceptable and non-acceptable welfare would need to be set very differently than how we might set this for those without Down syndrome. If what we are concerned about here is whether a future individual is likely to value their life and experience life as an overall positive experience, then this does not seem to justify the routinization of screening for this condition and the consequent recommendation for screening. The only way to set a threshold that might justify routinization and the impact this has on voluntary informed consent would be to set the threshold between acceptable and unacceptable quality of life based not on the expected welfare of an individual but on an external assessment of the value of this life.

## Does this mean we cannot justify any routine prenatal screening?

As we saw earlier, it is important, particularly with prenatal screening where a pregnancy involving a particular embryo has been established, to recognize that different ethical issues arise when choices do and do not impact the welfare of a future child. There will be types of routine screening that may allow us to improve the welfare of a particular future child. For instance, where screening identifies conditions such as syphilis or even HIV in a pregnant person, there are measures that can be taken that may reduce the harm done to the particular child who will be born. For these reasons, it may be that harm prevention could be used here to justify routine screening for such conditions. This would not be based on any threshold of acceptable harm risk but on the strong possibility that screening might prevent avoidable harm to a particular person who will be born. However, while there is more scope to justify routine screening based on harm prevention in these kinds of welfare-enhancing cases, before assuming that this will justify this approach, we should consider whether routine screening is the best course of action even in cases where harm may actually be prevented.[51] But given that Down syndrome cannot be prevented if we know about it during pregnancy, routinized screening for Down syndrome cannot be justified on the grounds of harm prevention. A child identified as having Down syndrome will either be born with this condition or not at all as a result of a decision to terminate the pregnancy.

It might be that we could justify routinization of prenatal screening for those rare conditions that are so severe that they are incompatible with what we would assess to be an intrinsically valuable life. We will explore this in more detail in the next chapter. But even if we were justified in routinizing screening for such extreme conditions, it is clear that Down syndrome is not one of these conditions that results in lives that are an overwhelmingly negative experience for those who experience them. In such cases, routine screening might be justified by harm prevention, as the quality of these lives would be so low as to be a harm to the person who lived that life, a life completely dominated by negative experience. However, (a) we would need to be open about the sort of conditions that would be likely to be above and below this threshold, to allow debate and transparency and (b) we would need to investigate whether routine screening, with its implications for influencing reproductive choices and sending negative messages about particular conditions is the most effective way to support pregnant people and to prevent these kinds of intrinsically harmful lives.

### *Where does this leave the case of routinization of screening for Down syndrome?*

It is extremely difficult to justify routine screening for Down syndrome on the basis that it empowers pregnant people and enhances their decision-making. There is consistent evidence that the usual standards for valid informed consent are not met in routine screening programmes of this kind, and the routinization of the screening arguably makes it fundamentally incompatible with the usual standards of voluntary informed consent. Further, offering screening and testing for Down syndrome in a way that is more compatible with how we offer other genetic tests would seem to have the potential to meet this goal of empowering pregnant people better, given that offering it in this way may allow the offer to be non-directive and remove the implicit recommendation that routinization of screening implies.

Those who are committed to the idea of routinization of screening for this condition could bite the bullet and attempt to justify this approach based on public health goals. But if this were to be done, then a strong argument would need to be provided as to why we should target Down syndrome as a condition that we are justified in aiming to eradicate. Given that pregnant people could still access screening and testing for this condition by electing to have this screen and test, we cannot make this argument based on the harm that might be done to those who wish to have information during pregnancy about this condition, either to allow them to end the pregnancy or prepare for the birth of this child. Pregnant people would still be able to access screening and testing if this was important to them, without routinization. Thus, the only justification for routinization here would be increased uptake of screening and thus termination of pregnancy to prevent the 'burden' of those with this condition on society more generally. Not only would any argument that tried to justify routinization on this basis be arguing for state-sanctioned eugenics, like the eugenics of the past, it is very difficult to understand where the justification for this could come from given that those living with this condition are as likely as anyone else to have a life they value.

In the absence of any other justification, what is left is a justification based on a harm threshold that simply doesn't work. A harm threshold approach would need to argue that those with Down syndrome have an unacceptable quality of life. We know from the testimony of those living with this condition that this is not the case and that people living with Down syndrome are as likely as any of us to value their lives. Our only other option here is to assess the lives of those with Down syndrome as unacceptable based on a comparison with other lives or another external factor. However, any attempt to assess the value of others' lives based on anything but the experience of those living these lives will be difficult to defend and again face accusations of state-sanctioned eugenics.

I argue that what may be happening here is that our unconscious negative assessments of disability in others, entrenched social norms around disability and the fact that routine prenatal screening for Down syndrome has been an established part of healthcare for over 60 years means that we have not questioned this approach to prenatal screening in a way that perhaps we should. Those of us who have not had personal experience of knowing someone with Down syndrome may assume, for all the reasons above, that having this condition is to suffer and that having this condition is bad for those who are born with Down syndrome. As a result, when we think about harm thresholds with regard to Down syndrome, these established policies, social norms and peculiarities of our unconscious mean that many of us automatically and unconsciously place those with Down syndrome below the threshold of what we consider to be an acceptable quality of life. However, while screening, testing and termination of pregnancy might well be the choice of many pregnant individuals, assuming that this is the case for everyone based on highly flawed ways of thinking about disability cannot be justified. Routine screening for Down syndrome reinforces unjustly negative assessments of people's lives and in some cases manipulates individual reproductive choices. When we recognize the social conditioning effect of established regulation and social norms and our unconscious responses to conditions such as Down syndrome, we can see that the justification we need for these infringements of individuals' interests is just not there.

**All three areas of regulation risk harming existing individuals by overriding their reproductive choices and stigmatizing particular groups.**

## Routine screening for Down syndrome

I have argued that routinization of screening for Down syndrome means that this screening has a directive approach and as such is incompatible for many pregnant people with the usual standards of voluntary informed consent. As a result, routine screening for Down syndrome runs the risk of influencing a significant number of people's reproductive choices without a robust argument to justify this approach.

But in addition to this, routinized screening for Down syndrome reinforces the unjustified social norms around this condition. I would support any pregnant person's decision to elect to access available screening and testing for Down syndrome and their ability to access termination of pregnancy. However, the routinization of screening is a very value-laden way of offering this screening and testing that implies a recommendation to accept screening. As we have seen, the main justification for the routinization of this screening is that having Down syndrome is somehow harmful to either the individual with the condition or the society they will live in. This justification, therefore, necessarily puts an unjustifiably low value on the lives of those living with Down syndrome – unjustifiable because this does not reflect the lived experience of those with this condition. This unjustifiably negative view of this condition reflects and reinforces stigmatizing attitudes to Down syndrome and those who live with this condition.

### *Pre-implantation genetic testing*

Having an explicit law that prohibits using pre-implantation genetic testing to either deliberately select for a condition like deafness or achondroplasia and other conditions viewed as 'serious', unless no 'unaffected' embryos are available, also has the potential to harm existing people and reinforce existing unjustified social norms around disability. Of course, most people would not use pre-implantation genetic testing to select for a condition that is widely seen as disabling, but not allowing this choice is highly problematic. As we have seen, it is difficult to make a case that allowing someone to choose to bring to birth a 'deaf' embryo harms that person. People who are born deaf are as likely as anyone else to value their life and will be born in the only condition they can be born into. It is not clear, therefore, on what grounds we should interfere with individual reproductive choices and prohibit the very small number of cases where an affected embryo might be preferred. In such cases, if we uphold this reproductive choice and allow the creation of a disabled life, we have allowed the creation of a life that is as likely as anyone else's to be valued. As a result, I argue that we do not harm anyone by making this choice.

It is likely that very few individuals would use pre-implantation genetic testing in order to select for a condition like deafness. However, there may be some individuals for whom at least having this choice is important. The existence of IVF itself is testimony to the fact that prospective parents will go to great lengths to attempt to have a child who is *like them*. Often gamete donation or adoption might be an easier, safer and, for those who need to fund this themselves, a cheaper option when it comes to having children, but this route is often seen as second best to having a child with our own genetic make-up. Where gamete donors are used, care is often taken to match this donor to the parents' appearance, ethnicity and other characteristics so that the resulting child will be as alike its parent/s as possible. Having a child who is a 'chip off the old block' or a 'mini me' and so on is something that people find fascinating and valuable. While there are, of course, many highly successful family units that are not genetically related and may not share many physical characteristics, having children who are like us, both physically, mentally, educationally, temperamentally and so on is something that is often very important to parents. That a deaf individual or couple might want to choose out of possible embryos to implant one that is *like them*, a deaf child, or that an individual or couple with achondroplasia might want to make a similar choice seems to fit very well with this general desire to create children like ourselves. Legally prohibiting this choice, as is the situation in the UK, or simply not allowing this choice as is the situation elsewhere, when there is little justification for doing so is highly problematic.

As we have seen, at least in cases of conditions such as deafness and achondroplasia, we cannot justify this prohibition based on harm avoidance. However, the language used in this UK regulation around pre-implantation genetic testing implies the opposite. The regulation talks about 'abnormality' rather than 'condition' and 'risk' rather than 'probability' and as such seems to imply that the lives it seeks to avoid are ones of negative value. As this is certainly not the case with many of the conditions this regulation addresses, this seems to reflect and enforce bias around disability and sends a very negative message to those living with the conditions that this regulation seeks to prevent.

In response, it might be argued that this regulation does not pass judgement on conditions such as deafness or achondroplasia that may be compatible with a high quality of life but is a way of allowing prospective parents to avoid conditions that they view as negative and thus to enhance reproductive choices. However, we know that this clause was introduced specifically to prevent this technology being used to deliberately choose to bring to birth deaf children, and if this really is about prioritizing reproductive choice, which I agree is an important goal, then a choice to choose differently, where this is not likely to create a life that is intrinsically harmful, must also be a legitimate choice.

A further counterargument to my position here might be that we cannot use scarce financial resources to fund choices that are generally viewed as unwise and that are likely to result in greater cost to the taxpayer in terms of the extra support that may be needed to bring up a child with a disabling condition. There is evidence that the introduction of routine screening for Down syndrome was originally justified on the basis of the cost savings that were likely to be made through 'successful' screening, which would reduce the costs of extra support needed for those born with this condition.[51]

There are all sorts of ways I could respond to this argument. I could argue that we do not use financial constraints to limit the number of children that people have more generally, even though each child will represent a cost in terms of education, healthcare and other public goods and there is always a risk with any reproduction that a child born will have additional needs. I could also argue that we do not, at least explicitly, prohibit people from reproducing who have a high risk of passing on a genetic condition that may create challenges for their children or require such individuals to access pre-implantation genetic testing. But even if you are not convinced by these responses, we could simply stipulate that these choices would not be prohibited but not publicly funded. Individuals could access private pre-implantation genetic testing to attempt to enable their reproductive choices to have children *like them*. At the moment, in the UK at least, implanting a deaf embryo after pre-implantation genetic testing is a legally prohibited act whether this is publicly or privately funded.

### Access to fertility treatment

While it seems that currently very few people are refused fertility treatment based on the Welfare of the Child considerations, every year, there are a significant number who do raise concerns under this provision.[53] Exactly how many is difficult to say, but there are enough to warrant the existence of ethics committees or groups attached to some clinics to consider these cases regularly.[54] I also argue that even if a patient is not explicitly refused fertility treatment, the process of investigation (which can take a significant amount of time possibly years in some cases[55]) is in most cases unnecessary (see the next section on the Intrinsic Value Position), stressful, risks stigmatizing particular groups and will lead to some patients becoming ineligible due to age, giving up completely or attempting to access treatment in another clinic (perhaps moving from a publicly funded provider to a private provider). A nurse in a UK clinic explains the typical process involved in these investigations saying,

> Unfortunately, the people that tend to come back with issues raised on the welfare of the child assessment tend to then have to undergo quite a – it can be quite an intense and invasive privacy process in trying to unravel the truth behind the issues the GP [General Practitioner] raises. Quite often there's something in the notes saying, I don't know, saying, 'Child fostered', 'Son fostered 1984' – or something like that. And the next step will be for one of the counselling team to see that couple or that person to talk to them about that and find out more about it.[56]

Cases where it is thought there are reasons for concern under the Welfare of the Child provision are typically discussed at length at a team meeting and in many cases taken to a dedicated ethics committee or a hospital ethics committee for further deliberation. This process may involve discussion by healthcare professionals and lay members of ethics committees and is likely to involve a great deal of information gathering from the prospective patients and other agencies.

This process seems to generally discriminate against those with fertility issues. Those without fertility issues are not scrutinized in this way when embarking on a pregnancy. Further, it is a particular subset of patients who will be scrutinized in this way – either those who volunteer information that is seen as a 'red flag' or those where these 'red flags' are conspicuous. So those with perceptible physical and mental health conditions are more likely to be scrutinized than those whose conditions are less obvious. Further, our sense of how serious these 'red flags' are perceived to be may well be influenced by biases around social class, racial background or other factors.

## Conclusion

I have argued that current regulation based on concern for the welfare of the child in the area of access to fertility treatment and use of pre-implantation genetic testing encourages decision-making based on subjectivity and has the potential to allow bias and unjustly negative impressions of disability to flourish. Further, these regulations do little to protect the welfare of any future children. The routinization of prenatal screening for Down syndrome is equally problematic. This approach to screening is either based on public health goals, which in this context appear closely aligned to state-sanctioned eugenics, or is based on an assessment of the quality of life of those with Down syndrome which is completely inaccurate and unfairly reinforces the stigmatizing negative perceptions of this condition and those who experience it. But what are the alternatives here and would they fare any better under the sort of scrutiny I have just subjected the current regulation to?

# CHAPTER 8
## ALTERNATIVE APPROACHES TO CURRENT REGULATIONS

I have argued that if we are to take seriously this commitment to be protective of the welfare of future children, then our current regulations cannot be justified. These current regulations are often based on and encourage unjustified unconscious negative assessments of others' lives. They also undermine individual reproductive choices unjustifiably and are ineffective in protecting the welfare of future lives.

While our current regulations in this area may be doing more harm than good, we also recognize that there will be some instances, where there is a risk of producing a child who is likely to have a intrinsically harmful life, where concern about the welfare of a future life may be an important consideration. If our existing regulation cannot be justified, how can we take our concern for the welfare of future children seriously without falling into the pitfall of subjectivity, bias and overly negative and stigmatizing approaches to this issue?

## Intrinsic Value Approach

During the exploration of these issues in this book, I have picked apart the arguments and concepts that underlie this idea of the welfare of future children. I have suggested that we often ask the wrong questions when we think about decisions about whether to bring to birth a particular future child, often asking questions about comparisons that cannot be made. I have argued that the only relevant question here is, 'Do we do something bad or harmful to this particular person by allowing them to exist with this particular condition or in these particular circumstances?'.

This question not only removes the temptation to make comparisons where comparisons are not applicable, but it also keeps the focus of our assessment firmly on the experience of the person who will live this life.

While it may seem obvious that if we are concerned about the welfare of future people what we should be concerned about is how *they* (the future people) may experience *their* life, whether this is likely to be an overall positive or negative experience, as we have seen, some positions make external judgements about the quality of future lives. They do this from a well-meaning drive to protect future individuals by claiming that we should strive to bring to birth the 'best' child possible or a child with a minimally decent quality of life. However, when we make decisions about which and whether to bring to birth future lives, by doing so based on anything other than how we expect that person to experience their life, we impose other people's standards of what makes a good life onto these future lives and move the focus of these decisions away from the welfare of individuals and towards one person or one group's idea of what makes a good life.

The position I take on where and why we should draw the line between what we consider to be an acceptable and unacceptable quality of life, I have called the Intrinsic Value Position. It is based on the idea that life is generally a good thing. As we have seen, those with Antinatalist positions may argue against this assumption. I think many of the Antinatalist arguments have some merit but, my experience of my life and what appears to be the experience of the majority of human beings, is that we value our lives highly as a positive experience.[1] My Intrinsic Value Position focuses very much on the lived experience of individuals. It starts with this assumption that life is generally a good thing and aims to mitigate the effects of subjectivity, bias and the effect of social norms in making welfare assessments of future lives by focusing firmly on learning from the experience of individuals who live with the sort of conditions and disadvantages that we are considering.

I argue that this is the only justifiable approach when attempting to make assessments of future lives and thus the only acceptable way to draw this threshold between what are expected to be acceptable and unacceptable qualities of life is to base this threshold on the lived experiences of those experiencing similar lives. It is only by focusing on the lived experience of actual individuals that we can make an informed assessment of whether a life is likely to be one that is intrinsically valuable or intrinsically harmful to that individual.

I argue that the adoption of this Intrinsic Value Threshold would maximize the reproductive autonomy of prospective parents. This threshold does not suggest that we have any duty to bring to birth any child and emphasizes reproductive autonomy, so that parents who do not wish to bring to birth a child with a condition such as deafness or Down syndrome should be supported to access screening and testing and, where chosen, termination of pregnancy to enable these choices. The only possible justification for limitations to reproductive choice would be where there is a significant risk of a child being born with a life so severely compromised by negative conditions that this life is likely to be intrinsically harmful to the individual experiencing it. Exactly where this threshold will be set and what kinds of lives would be assessed to be in this category of intrinsically harmful lives below the threshold will be something that will take significant research into the lived experience of those with these kinds of extreme conditions.

This will still not be a perfect foundation for regulation in this area. Like many ethical questions, there is probably no perfect answer to this question of how we safeguard the welfare of future lives in an ethically justifiable way. Assessments of future lives made on this Intrinsic Value Position will turn out to be inaccurate in some cases. However, what we currently have is regulation that gives little in the way of guidance around decision-making regarding the welfare of future children. This puts responsibility on healthcare professionals to make these decisions, but such wide and ambiguous guidance is used that it is ultimately left to the clinics or healthcare professionals to make these decisions. Explicitly adopting this Intrinsic Value Threshold and being prepared to explain why this threshold is adopted and to develop transparent and detailed regulation based on this threshold is, I argue, a much better and justifiable basis for regulation. This approach would allow us to provide clearer support for decision-making, enable much greater consistency and accountability, minimize the scope for the influence of bias and prejudice and avoid the stigmatization of particular conditions and social circumstances.

If we were to develop regulation based on the Intrinsic Value Position, this would have a significant impact on regulation in this area.

### The Intrinsic Value Position as applied to access to fertility treatment

If we apply this threshold to the question of access to fertility treatment, it would only justify investigation and possible intervention in cases where we have good reason to believe that the resulting child would have a life that is completely dominated by suffering.

Under the application of this threshold, most of the cases that we are currently concerned with under the Welfare of the Child provision when it comes to access to fertility treatment, would not trigger further investigation. This would mean that something like alcohol or drug use would only be of concern if there was evidence that this would be so serious as to overshadow the positives of this child's life. Predicting the future drug and alcohol use of a person and its effect on a child is something that is based on guesswork, on the information that is provided to you (which may not be accurate) and subject to change (it may be that someone will seek help if they are expecting a child and this might be even more likely if individuals do not feel they need to conceal these issues from healthcare professionals). Physical disability and mental health conditions are also unlikely to be something that would preclude treatment, particularly where support can be put in place, where needed, just as it would be where there is no fertility issue and conception occurs naturally. Other factors such as social conditions, criminal convictions, child protection issues and so on would only be considered relevant here if it were thought that the impact on a child would be likely to be so severe as to overwhelm the positives of that life. Intervention would then only be justified in extreme cases.

## *The Intrinsic Value Position as applied to pre-implantation genetic testing regulation*

Applying this threshold to pre-implantation genetic testing regulation would mean that it would only be justifiable to prohibit implanting embryos with conditions that are thought to be likely to cause a life to be intrinsically harmful. This would *not* apply to conditions like deafness or many of the other conditions that are currently prohibited to be preferred under UK regulation.

In deference to reproductive autonomy, individuals who have a genetic risk of having a child with a condition they viewed as negative, including deafness, would be able to access pre-implantation genetic testing to enable their reproductive choices. However, at the same time, individuals with conditions like deafness or other conditions often seen as disabling, who wished to use pre-implantation genetic testing either as part of IVF or as a stand-alone service to try and have a child like them would not be prohibited from doing so, so long as the condition in question was not likely to impact on that life so dramatically as to be likely to make it intrinsically harmful to the person who experiences it. It is likely that very few prospective parents will choose in this way[2] but not prohibiting this choice is important for the reasons set out above.

## *The Intrinsic Value Position as applied to routine screening for Down syndrome*

Based on the Intrinsic Value Position, we do not have good reasons to make screening for Down syndrome a routine part of prenatal care. We know that this routinization can have a significant effect on the voluntariness of the consent given to screening. While in some situations prevention of harm may justify this unusual approach to consent, when our focus is squarely on the welfare of future children, this does not give us the evidence of harm needed to justify routinization. We know that Down syndrome is a condition that is not any more likely to result in what might be considered intrinsically harmful lives, that is, lives that are dominated by negative experience, than an absence of the condition.

I argue that what is important here is respect for reproductive choice and reproductive autonomy. Access to balanced and accurate information about screening for Down syndrome and the condition itself should be provided in a way that enables pregnant people to make a decision as to whether they wish to elect to be part of screening for this condition. A great deal of consideration should be given to how this information can be provided in a way that mitigates the effect of unconscious bias and unjustly negative attitudes towards this condition and disabilities more generally. Work should be done to ensure that pregnant people understand that screening is a choice and not accepting screening is just as responsible an option as accepting screening. This might mean doing work to attempt to tackle the social norms around this condition, and removing routinization of screening for this condition would be an important part of this process.

Currently, the routine screening for Down syndrome is done at the same time as other routine screens for chromosomal disorders – Patau and Edward Syndrome. These conditions are usually more impactful in terms of challenges for those who live with them and many affected foetuses will not survive to birth.[3] Whether these conditions fall below the Intrinsic Value Threshold is something that would need to be explored in some detail, but there is nothing in this approach that would prohibit the justification of routine screening for conditions that were assessed to be likely to fall below this threshold. What would be important here is that the justification for routine screening for certain conditions must be clearly set out so that we understand why this different approach is taken for some conditions. Being clearer and more transparent about routine screening in cases where there is a high risk of creating intrinsically harmful lives would allow debate on this issue and make this policy more easily challengeable.

### *What are the benefits of this Intrinsic Value Approach to regulation in this area?*

- This approach would mean that regulation could be based on a threshold that could be explicitly outlined. In doing so we can be transparent about the reasons for setting the threshold where we do and allow debate about its acceptability.

- Setting this threshold would vastly reduce the amount of subjectivity that is currently involved in decision-making in this area and thus increase consistency of decision-making (in clinics and nationally) and mitigate some of the influence of bias and prejudice.

- Adopting such a threshold would mean we have a much greater chance of applying regulation consistently and transparently. Under this threshold, we could provide examples of what conditions do and do not fall below the threshold here, and this could then be applied in every case.

- This approach would reduce the number of conditions that are viewed as 'serious' risks of harm in a way that we can justify and explain. This would reduce the amount of stigmatization around certain conditions and may even begin to change societal bias against these conditions.

- This approach maximizes reproductive autonomy, allowing those who wish to avoid conditions they view as negative to do so but only limiting reproductive choices in the rare cases where individuals might want to select for a condition or reproduce in conditions that are likely to cause the resulting life to be an intrinsic harm to the person who experiences it.

- This threshold also recognizes and learns from the lived experience of those with particular conditions taking their experience of these conditions as the reason to allow or restrict reproductive choices.

- The explicit setting of a threshold that can be explained and justified would provide a strong foundation for further regulation in this area. With a vast number of new pre-implantation and prenatal tests on the horizon, having a workable, justifiable position regarding the welfare of future children will allow regulation to be developed that is consistent, transparent, practically useful in decision-making and rests on sound ethical reasoning.

- When it comes to access to fertility treatment, removing the threat of refusal of treatment may enable individuals and couples to be more open about any challenges they may face regarding their own social, psychological or physical situation when attending a fertility clinic. This might present a valuable opportunity to provide support that may improve their own lives and the lives of the family and any future children they may have.

- Reducing the scope of concern about the welfare of the child may help to remove some of the inequality of scrutinizing those who need assistance with reproduction. Individuals and couples who do not need assistance to reproduce are not subjected to scrutiny about their ability to provide a child with what is considered to be acceptable family conditions. Reducing the scope for this scrutiny significantly would allow us to minimize this inequality for prospective parents who, due to infertility, sexuality or relationship status, wish to use fertility services.

Of course, while this approach does have benefits, I can imagine many objections to this way of regulating in this area. In line with our ARC approach, it is important that I identify these and deal with them if I want to defend my position and hold it with any confidence. So what might these objections be?

### This approach has not removed the grey area of subjectivity

While applying the Intrinsic Value Threshold will not remove the grey area of subjectivity when it comes to assessments of the welfare of future children completely, it will reduce the scope of this subjectivity. Any attempts to make judgements about other people's quality of life when these lives have not yet come into existence will be unavoidably subjective. However, the scope of this subjectivity can be reduced by taking certain measures. With a clearer, less ambiguous basis for this threshold, it will be possible to consult with those living with different conditions to provide guidance and examples as to which conditions we should be concerned about and which conditions do not fall within the remit of these regulations. There will be no way of completely avoiding bias, ambiguity and inconsistency in the way these decisions are made. But with a focus on learning from the lived experience of existing people, it is likely we can improve the clarity, transparency, robustness and consistency of these decisions.

The main difficulty with applying this threshold is deciding which conditions fall below the threshold – that is, which conditions are likely to result in a life so dominated by negative experience that it is likely to be considered a harm by the person who experiences it. While this is a difficulty, I suggest it is still a huge improvement on current approaches where *any* condition could be classified as a 'serious disability', with no ability to appeal these decisions. While it is, of course, possible that on the Intrinsic Value Threshold, mistakes in classifying something as above or below the threshold could occur, the narrowing of the criteria would allow these decisions to be appealed and allow us to revise guidance based on this much more transparent and workable approach.

### Isn't this just rejecting the idea of concern about the welfare of future children?

Basing regulation on the Intrinsic Value Threshold will mean that, in the case of access to fertility treatment, we would not need to 'investigate' as many individuals seeking treatment regarding our concerns about the welfare of the children they may have. The explicit adoption of this threshold would allow guidance to be clearer about where this threshold lies and why and what sort of cases might lie on either side of this threshold. Crucially, it would be possible to be clearer about which sort of cases do *not* fall below this threshold – for example, deafness or Down syndrome and so on – and where parents will need additional but available support.

Similarly, when this threshold is applied to pre-implantation genetic testing, there will be a great many conditions that no longer fall into the category of 'serious' harms, again including deafness, Down syndrome and many other conditions that are currently considered unacceptable to be preferred to non-affected embryos. Thus, to maximize reproductive choice, while individuals would be able to access pre-implantation genetic testing in an effort to avoid these conditions, there would be no prohibition on choosing in favour of these conditions. In this way pre-implantation genetic testing could be used much more effectively to empower prospective parents with information to enhance their reproductive choices whatever those choices might be.

However, these changes do not mean that regulation based on this revised threshold does not show concern for the welfare of future children.

The Intrinsic Value Threshold is focused entirely on the welfare of future children and the idea that there will be lives that are intrinsically harmful to those who experience them and that these lives should be avoided where possible.

### *There is a danger that this threshold will allow intrinsically harmful lives to be born*

There will always be a danger that applying this threshold will allow some intrinsically harmful lives to be born. However, the danger that this will happen is likely to be rare, and any threshold that we apply will face the same risks. For instance, it is impossible to predict from the limited contact that assisted reproduction clinics have with patients what challenges these families will face moving forward. There will be some seemingly 'unproblematic' individuals and couples who will unexpectedly face physical, psychological, social and financial challenges that were not foreseen at the time of treatment. There will be prospective parents who were able to mask or were in denial of the challenges they face, which thus were not apparent to the clinic.

It has been argued that in the UK, the Welfare of the Child provision was motivated by the 'spectre of the paedophile' that might use fertility treatment to create their victim.[4] While of course, this will be a possibility under regulation and guidance based on the Intrinsic Value Threshold, this is currently also a possibility, and moving this threshold, while more justifiable on other grounds, is unlikely to change this risk significantly.

Applying this threshold to the use of pre-implantation genetic testing will risk the creation of lives that fall below this Intrinsic Value Threshold where information about a condition is perhaps not detailed enough or a condition is combined with other challenges. However, the current approach to pre-implantation genetic testing will not be effective in avoiding the creation of intrinsically harmful lives. Embryos selected as part of pre-implantation genetic testing may well turn out to have intrinsically harmful lives for other reasons than the conditions that might be tested for. Allowing the prospective parent to use pre-implantation genetic testing to select for deafness or other conditions seen as disabling, will not affect this risk of creating intrinsically harmful lives.

Moving to a threshold that I argue is less discriminatory and more transparent, in terms of why this threshold has been chosen, will not change the fact that in some very rare cases, decisions will be made that result in a life that is intrinsically harmful for the child who lives that life.

## *Offering screening and testing for Down syndrome in a non-routinized way will undermine the quality of pregnant people's choices*

No longer making Down syndrome screening and testing a *routine* part of prenatal care and emphasizing voluntariness and removing pressure, where possible, to accept screening may reduce the numbers of pregnant people who opt for this screening. The main objection to this approach to prenatal screening and testing would likely be that it risks missing an opportunity to provide information for some pregnant people who would have wanted this information but for whatever reasons did not opt into screening or testing. This is a concern as it might be argued that having this information about the foetus you are carrying would allow you to make better and more informed choices about your pregnancy and your life. However, this argument could be put forward to justify pressure on individuals to accept all sorts of screening and testing. For instance, it could be argued that having information about genetic conditions such as Huntington disease (a serious inherited genetic condition where symptoms start between thirty and fifty years old and is significantly life-limiting) would be useful for those who will develop the condition or who may pass this condition to their children. However, out of respect for individual autonomy, we do not usually think it is justified to put pressure on people to accept testing for serious conditions like this, and in fact, there is evidence that those with a risk for this condition invariably do not wish to test for it early in their lives.[5]

If we really want to empower pregnant people's choices, then we should take seriously the idea of empowering all choices whatever they might be. At the moment, routinization of Down syndrome screening may empower the choices of those who feel strongly that they wish to know whether their foetus has this condition and perhaps wish to terminate their pregnancy. However, this routinization cannot be said to empower the choices of those who wish to decide whether having this information is important for them.

There will be those who argue that this alternative approach to screening and testing for Down syndrome would be too time-consuming and too expensive to implement. Spending time with pregnant people to explain the pros and cons of screening and to emphasize the voluntariness of this procedure is likely to take more time and cost in terms of training. However, if we are really serious about empowerment of pregnant people's choices, then we cannot continue to deny the evidence that routine screening for this condition is not the best way to achieve this.

### *I feel uncomfortable about this*

As we have seen, there is a shared intuition that choosing to bring to birth children who may face more challenges than other children is something we are not comfortable with. Many of us intuitively feel that this is the wrong thing to do. However, I have argued that basing regulation on intuition is not something that is generally a good idea. Doing so means that we are in danger of basing regulation that has the power to influence and override individuals' fundamental choices on intuitions that may turn out to arise from bias and even prejudice rather than being backed by reasons that we are able to discuss and defend. Unless we can provide good reasons as to why applying this threshold is not the right thing to do and why our previous approaches were more justifiable, then, it seems that we may have to accept, in this case, that our intuitions were not ones we want to give moral weight to.

### *I am a healthcare professional and feel I have a duty of care towards the children I help to bring to birth*

This final objection raises the issues we explored earlier about healthcare professionals feeling they have a particular obligation towards any children they were instrumental in helping to bring to birth.

While I argue that the Intrinsic Value Threshold is the preferable basis for regulation when it comes to the welfare of future individuals, I recognize that my role here is significantly different to that of healthcare professionals working, for instance, in fertility clinics who may feel a duty of care towards these new lives. As we saw earlier, this idea of a special duty of care to children that healthcare professionals helped to create is something that is widely acknowledged. For instance, in a study of the application of the Welfare of the Child provision in two UK clinics in 2002, Kathryn Ehrich et al. explain that clinic staff participating in the study 'expressed a heightened sense of responsibility' for the welfare of any child they were instrumental in bringing to birth, with one Embryologist saying that

> Obviously, whilst the average couple in the street want to have a child, then there's no laws against it. But then they don't use the intervention of a third party and actually to let me sleep at night, I'm quite glad, that, at least on a very basic level, that the welfare of the child aspect is there I think we are intervening, so I think there should be some sort of check.[6]

It was argued that for many staff, the welfare of the child assessment 'is closely bound up with their own sense of professional accountability'.[7] While there is evidence that this is not a universal feeling, most staff felt that the Welfare of the Child provision should be retained.[8]

The strong support that the Welfare of the Child provision has in the area of assisted reproduction was also apparent in a significantly larger study reported in 2014, where it was found that 'not a single interviewee argued for abolishing the WOC [Welfare of the Child] assessment'.[9]

Given this perceived duty of care or responsibility that healthcare professionals working in this field feel, there is likely to be opposition from healthcare professions to regulation based on the Intrinsic Value Threshold. The idea of only having a responsibility for avoiding the creation of lives that are predicted to be a significant risk of being intrinsically harmful to those who experience them is something that healthcare professionals, particularly those working in fertility clinics who are used to making assessments based on a much higher threshold, are likely to find unpalatable.

Is a compromise position when it comes to applying the notion of the welfare of the child to future children a viable option here? Can we, for instance, improve the current approach to this issue by being much more explicit about the thinking behind the threshold of 'serious' or 'significant' harm or disability. This might be done by providing much more detail about where this threshold lies, clarifying what criteria we might have for conditions and circumstances that fall below this threshold and putting in place mechanisms to minimize the effects of bias and the disability paradox. In doing so, the hope might be that we can identify a notion of 'serious' or 'significant' harm or disability that speaks to the general concerns about the welfare of future children, particularly where healthcare professionals are instrumental in their coming to birth, but that stands up better to scrutiny, is more helpful in practice and can be applied more consistently.

However, while it may be possible to make improvements to the way in which the existing criteria are applied, there is a fundamental problem with this approach and with the current regulation in this area that cannot be improved by clarifying what is meant by 'serious' or 'significant' harm or disability. This goes back to my concern about any harm threshold that measures things other than the expected experience of future individuals. If we assume that the Intrinsic Value Threshold would be unpalatable to healthcare professionals, this seems to imply that only a threshold *above* this one might recognize the duty of care that healthcare professionals feel they have when helping to create new life. However, as we have seen, having a threshold here that is based on external views of what is a minimally decent life and that measures something more than the expected experience of the person in question is very difficult to justify.

Training healthcare professionals to understand the effect of negative social norms and biases on their feelings of responsibility when creating new lives may make a criterion based on the Intrinsic Value Position more palatable for healthcare professionals. Being open and transparent with the reasons for the Intrinsic Value Approach and illustrating the problems with the current approaches may also help to ease any concerns that healthcare professionals might have about a change in approach. What is clear is that the current approach here, while popular with many healthcare professionals, is one that is not only ethically unjustifiable, but is also unhelpful and reinforces the social norms that amplify healthcare professionals' feelings of responsibility in this area. I argue that attempting to prop up existing regulation is not an option as it is based on flawed foundations.

## Moving away from regulation based on the idea of the welfare of future children?

I have argued that the current regulation used to guide decision-making and justify policy in the areas of assisted reproduction, the use of pre-implantation genetic testing and routine screening for Down Syndrome cannot be justified. Even if we try and improve the guidance provided for those making these decisions, these approaches are based on reasoning that does not stand up to scrutiny. If we really want to take the welfare of future children seriously then, I suggest, we need to base regulation in this area on something like the Intrinsic Value Position where the assessment of the welfare of future individuals is taken very seriously and a highly informed and self-reflective approach is developed to mitigate the effects of unconscious thinking and bias.

There is, of course, another option here and that is to remove the need to consider the welfare or best interest of a future child altogether in the context of access to assisted reproduction and pre-implantation genetic testing.[10]

### *Why might we consider this approach?*

If assessments based on the welfare of future children were removed in the context of assisted reproduction, all those who are medically and financially (either through public funding or private means) eligible for treatment would be treated. This would remove the highly subjective nature of the Welfare of the Child assessment and the stigmatizing, discriminatory and inconsistent application of this provision currently.

Prospective patients may even be more likely to divulge any physical, psychological or social issues that they feel will be a challenge and thus be able to receive support for these issues. Other individuals who do not have fertility issues, either due to medical issues, their sexuality or other social reasons, are not scrutinized in this way when they embark on pregnancy. In other contexts of healthcare, social judgements about the eligibility of individuals for treatment are usually not encouraged. As a result, it might be argued that requiring this assessment of those needing the help of fertility services is discriminatory. Without this need to consider the welfare of the resulting child, those who for whatever reason need help with fertility issues would then be treated more equitably with those who do not.

If this requirement to consider the welfare of future children was removed in the context of access to fertility treatment, child protection measures would then work in the same way for these individuals as it does for all of those who are able to reproduce without any medical assistance. Where there are concerns for the safety of existing children in actual circumstances that can more easily be assessed, then measures can be taken to support these families and, where necessary, to ensure these children's welfare.

This kind of intervention will only be necessary in relatively rare cases, many of which we could not predict at the pre-conception stage. Thus, even if we remove the need for the welfare of the child assessment before access to fertility treatment, it can be argued that we are still taking the welfare of future children seriously. It is extremely unlikely that removing this requirement would result in the creation of many, if any, additional children with lives that would be considered to be intrinsically harmful. Even with this welfare of the child assessment, there is, of course, always a danger that children will be born with these kinds of extremely challenging lives as a result of unknown or unforeseen circumstances. By preventing access to fertility treatment for some patients we

- do not make anyone's welfare better or worse as any child who would have been born would have been born in the only condition they could be, with the challenges we might be concerned about.

- We may not prevent the birth of a future child as this/these prospective parent/s may go elsewhere to be treated. They may find that other publicly funded or private clinics will not take the same approach, or they may not divulge the same information to another clinic, missing out on any support that could be offered for them and their family.

- Under the current regulation, there is already a risk that children will be born with intrinsically harmful lives as a result of undisclosed or unforeseen circumstances. Assessment based on the concern for the welfare of future children can never prevent this risk.

Removing this requirement when it comes to assisted reproduction and pre-implantation genetic testing would have a very limited impact on who is born. Very few people are refused fertility treatment using current regulations.[11] There is likely to be very little demand for the use of pre-implantation genetic testing to select for conditions such as deafness or achondroplasia.[12] Withdrawing this requirement in the context of pre-implantation genetic testing would allow deaf individuals and those with other conditions to attempt to have a child *like them*, a goal that is usually supported by assisted reproduction generally. It may also help to tackle unjustifiable negative social norms around disability rather than reinforcing them.

Routine screening for Down syndrome is not explicitly motivated or regulated by the idea of the welfare of future children. However, as we have seen, other motivations of routinization of this screening such as empowerment of pregnant people's choices and public health goals are very difficult to justify in this context. Removing the welfare of the child regulation in other areas when it comes to future lives would provide an opportunity to investigate the motivations of routinization of screening for Down syndrome in detail. If empowerment of pregnant people's choices is identified explicitly as the primary motivation for this policy, then ways of enabling this empowerment and the upholding of reproductive choices could be developed away from routinization.

A move to remove any requirements for regulation in this area to be motivated by assessments of the welfare of future children would be controversial. We know that our intuitions in this area are strongly in favour of attempting to protect future lives from conditions that we might see as disadvantageous. In line with this, we know that healthcare professionals, the public, policymakers and many high-profile academics feel strongly that considering the welfare of future children is paramount particularly in instances where the state or healthcare professionals are instrumental in helping an individual or couple to reproduce. However, criminalization of homosexuality and regulation that reinforced and enforced the inequality of women and people of colour also had great intuitive appeal to large sections of society. I argue that we must be vigilant here and make sure we do not make a similar mistake with regulation based on this notion of the welfare of the future child.

While overturning regulation in this area may be unpopular at first, it would enable debate in this area and transparency about the issues and problems that undie these regulations. It may enable more people to explore the impact that their unconscious thinking may have on their opinions on these issues and many other issues and encourage the questioning of widely held social norms and biases. It may also enable the reduction of stigmatization of individuals and groups in society. What is true is that if we do not address the fundamental problems of grounding regulation on concerns about the welfare of future children, we are complicit in upholding unjust regulations which limit the reproductive choices of individuals unfairly and reinforce and encourage biased social norms and attitudes around disability and particular social conditions.

## Conclusion

For over twenty-five years, I have argued that this notion of the welfare of future children has been used as a smokescreen behind which bias and prejudice flourishes.[13] In writing this book, I hope I have been able to provide the reasons for this claim. The questions, principles and concepts that underlie this approach to regulation in this area are complex and often impenetrable to those from outside this area of discourse. My aim with this book was to blow away this smokescreen and lay bare the issues, complexities and difficulties of regulation based on the welfare of future children in an area that is so influenced by intuition and unconscious thinking. In doing so, I hope to ignite the debate in this area and allow us to have the courage to address our own biases and the entrenched social norms of our society to come to a position on these questions that we can be confident aligns with our core values of respecting the welfare, interests and choice of individuals. I have argued that we must protect individuals from the consequences of allowing other's ideas of what constitutes a good enough quality of life to infringe on their freedom to make the choices that are right for them.

I want to reiterate here that my arguments focus firmly on this idea of respecting individual reproductive choices. As a result, I wish to emphasize that nothing I have said in this book implies that we have a moral obligation to bring to birth children with disabilities or with or in other conditions that are thought to be disadvantageous. I argue that people should be able to choose whether they continue a pregnancy or whether they are able to access pre-implantation genetic testing to avoid selecting an embryo with a condition, or whether they wish to choose to have their foetus screened for Down syndrome. However, if our commitment to choice, and reproductive choice in particular, is one that we are serious about and serious about applying fairly to all members of the population, then we should allow those who wish to make other choices when it comes to bringing to birth a child with these kinds of conditions or in conditions that others regard as challenging, to do so.

I started this book asking whether it might be possible that our regulations around access to fertility treatment, use of pre-implantation genetic testing and routine screening for Down syndrome are based on bias and prejudice rather than reason and compassion. I asked whether it might be possible that regulation based on concern for the welfare of future children is not only ineffective but unjustifiably discriminatory, encouraging decisions based on subjective and often biased judgements.

I have answered these questions in detail by outlining and analysing the central arguments and issues in this debate. I have presented an overview of regulation in this area, analysis of the bioethical and philosophical arguments and concepts that underlie these regulations, explored psychological insights around unconscious decision-making and suggested possible alternative approaches to these complex ethical issues. In doing so, I have argued that that these widespread regulations that appear to be founded on notions of compassion and protective instincts towards future children are unjust and unjustifiable. Ultimately, I argue that we have a duty to repeal these regulations or be guilty of enabling the unfair treatment of individuals and the amplifying of inaccurate negative social norms and biases that harm many more.

Of course, you might not agree with the conclusions that I have come to here. As I stated at the start of this book the point of this book was not necessarily to get you to agree with my position on these issues. My aim was to enable you to explore your own position in some detail, to examine your own reasons for the positions you take and to take a much more questioning approach to the issues that we have explored. I have provided what I hope is a clear exposition of my arguments and the counterarguments to these arguments, to enable you to explore these issues and defend your position. Coming to a defendable position on regulation about the welfare of future human lives is important, not only because of the individual lives that may be directly affected by these regulations but because of the wider consequences of these regulations. Whether you agree with my conclusions or not, being open to consider this alternative view of regulation based on the welfare of future children will mean we can open up this debate and work towards establishing regulation in this area that we can be confident in supporting.

# NOTES

## INTRODUCTION

1   This book focuses on the ethical evaluation of legal, regulatory, policy, guidance and professional provisions that apply to the area of reproduction both in the UK and internationally. For the sake of brevity and ease of reading, I will use the terms 'regulation' and 'regulatory framework' to denote these various ways in which reproductive choices are governed.

2   See, for example, E. E. Andrews, R. M. Powell and K. Ayers, 'The evolution of disability language: Choosing terms to describe disability', *Disability and Health Journal* 15(3) 101328 (2022): 1.

3   Andrews, Powell and Ayers, 'The evolution of disability language', 2.

4   Andrews, Powell and Ayers, 'The evolution of disability language', 2.

5   Andrews, Powell and Ayers, 'The evolution of disability language', 3.

6   While there are some that see an important difference between the terms 'moral' and 'ethical' my understanding is that Harris did not and I have taken his lead on this. Therefore, in this book the terms 'moral' and 'ethical' will be used interchangeably and synonymously.

7   J. Savulescu, 'Procreative beneficence: Why we should select the best children', *Bioethics* 15(5/6) (2001): 413–26.

8   R. Bennett, 'When intuition is not enough', *Bioethics* 28 (2014): 447–55.

9   V. A. Kushnir, G. D. Smith and E. Y. Adashi, 'The future of IVF: The new normal in human reproduction', *Reproductive Sciences* 29 (2022): 853.

10   In this book I have, wherever possible, used the inclusive language of pregnant people rather than pregnant woman or women. I do this to recognize that many pregnant people will not identify as women. However, many of the outside sources I refer to use the language of women and pregnant women to describe the same population. For the sake of accuracy, I have not altered these outside sources.

11   R. Winand, K. Hens, W. Dondorp, G. de Wert, Y. Moreau, J. R. Vermeesch, I. Liebaers and J. Aerts, '*In vitro* screening of embryos by whole-genome sequencing: Now, in the future or never?', *Human Reproduction* 29(4) (2014): 842–51; N. M. Murphy, T. S. Samarasekera, L. Macaskill et al., 'Genome sequencing of human *in vitro* fertilisation embryos for pathogenic variation screening', *Scientific Reports* 10 (2020): 3795.

12   H. Bowman-Smart, C. Gyngell, C. Mand, D. J. Amor, M. B. Delatycki and J. Savulescu, 'Non-invasive prenatal testing for "non-medical" traits: Ensuring consistency in ethical decision-making', *American Journal of Bioethics* 23(3) (2023): 3–20.

## Notes

13   For further information, see, for example, F. Ulph and R. Bennett, 'Psychological and ethical challenges of introducing whole genome sequencing into routine newborn screening: Lessons learned from existing newborn screening', *The New Bioethics* 29(1) (2022): 52–74.

14   To be completely accurate here, rather than saying 'implant a "deaf" embryo' I should say 'have a "deaf" embryo transferred' as the actual implantation of an embryo into the uterus after it is transferred into a recipient is not something that can be controlled. However, I have chosen to use the word 'implant' to clearly indicate the intention to bring this embryo to birth which might be less clear to a non-clinical audience if the word 'transferred' was used.

15   See, for example, R. Bennett and J. Harris, 'Restoring natural function: Access to infertility treatment using donated gametes', *Human Fertility* 2 (1999): 18–21; E. Jackson, 'Conception and the irrelevance of the welfare principle', *Modern Law Review* 65(2) (2002): 176–203: S. Waxman, 'Applying the preconception welfare principle and the harm threshold: Doing more harm than good?', *Medical Law International* 17(3) (2017): 134–57.

16   See, for example, D. Benatar, *Better Never to Have Been: The Harm of Coming into Existence* (Oxford: Oxford Academic, 2006); M. Häyry, 'A rational cure for prereproductive stress syndrome', *Journal of Medical Ethics* 30 (2004): 377–8.

17   I use the term 'service user' in this context instead of alternatives such as patients to acknowledge that these individuals are not necessarily ill, and patient may not fit. However, I also recognize the limitations and criticisms of the term 'service user' in this context.

## CHAPTER 1

1   United Nations (1989) *Convention on the Rights of the Child* (New York: United Nations).

2   For more information on this issue of moral status, see, for example: B. Steinbock and P. T. Menzel. *Bioethics: What Everyone Needs to Know?* (Oxford: Oxford University Press, Incorporated, 2023) Chapter 7; A. Jaworska and J. Tannenbaum, 'The Grounds of Moral Status', *The Stanford Encyclopedia of Philosophy* (Spring 2023 Edition), E. N. Zalta and U. Nodelman (eds), https://plato.stanford.edu/archives/spr2023/entries/grounds-moral-status/.

3   See, for example, HFEA, *Tomorrow's Children* (HFEA, November 2005), 6, 7, 14.

4   J. Harris, 'One principle and three fallacies of disability studies', *Journal of Medical Ethics* 27 (2001): 383–7; J. Savulescu, 'Procreative beneficence: Why we should select the best children', *Bioethics* 15 (5/6) (2001): 413–26; J. Glover, 'Future People, Disability, and Screening', in P. Laslett and J. Fishkin (eds), *Justice between Age Groups and Generations* (New Haven: Yale University Press, 1992), 127–14, [reprinted in John Harris, ed. 2001. *Bioethics* (Oxford: Oxford University Press), 429–44].

5    International Federation of Fertility Societies' Surveillance (IFFS), 'Global trends in reproductive policy and practice', 8th edition. *Global Reproductive Health* 4(1) (2019): e29, March 2019.

6    G. Pennings, 'The welfare of the child: Measuring the welfare of the child', *Human Reproduction* 14(5) (1999): 1146.

7    S. Franklin, 'The HFEA in context', *Reproductive BioMedicine Online* 26(4) (2013): 310.

8    The HFEA code of Practice, 9th edition, available online: https://portal.hfea.gov.uk/knowledge-base/read-the-code-of-practice/.

9    The HFEA code of Practice, 9th edition, Section 8.14 (iv), available online: https://portal.hfea.gov.uk/knowledge-base/read-the-code-of-practice/.

10   S. Golombok, 'The psychological wellbeing of ART children: What have we learned from 40 years of research?' *Reproductive Biomedicine Online* 41(4) (2020): 743–6.

11   S. Sheldon, E. Lee and J. Macvarish, 'Supportive parenting, responsibility and regulation: The welfare assessment under the reformed human fertilisation and embryology act' (1990), *Modern Law Review* 78(3) (2015): 472.

12   Sheldon, Lee and Macvarish, 'Supportive parenting, responsibility and regulation', 461–92.

13   Sheldon, Lee and Macvarish, 'Supportive parenting, responsibility and regulation', 473.

14   S. L. de Lacey, K. Peterson and J. McMillan, 'Child interests in assisted reproductive technology: How is the welfare principle applied in practice?', *Human Reproduction* 30(3) (2015): 620.

15   Sheldon, Lee and Macvarish, 'Supportive parenting, responsibility and regulation', 477.

16   International Federation of Fertility Societies' Surveillance (IFFS), 'Global trends in reproductive policy and practice', 8th edition. *Global Reproductive Health* 4(1): e29, March 2019: 99

17   International Federation of Fertility Societies' Surveillance (IFFS), 'Global trends in reproductive policy and practice', 9th edition. *Global Reproductive Health* 7(3): e58, Autumn 2022: 131.

18   International Federation of Fertility Societies' Surveillance (IFFS), 'Global trends in reproductive policy and practice', e58, Autumn 2022: 130.

19   National Health and Medical Research Council, *Ethical Guidelines on the Use of Assisted Reproductive Technology in Clinical Practice and Research* (Australian Government, updated in 2023), accessed online: https://www.nhmrc.gov.au/about-us/publications/art.

20   https://laws-lois.justice.gc.ca/eng/acts/A-13.4/.

21   V. Fournier, D. Berthiau, J. d'Haussy et al., 'Access to assisted reproductive technologies in France: The emergence of the patients' voice', *Med Health Care and Philos* 16 (2013): 56.

22   Nordic Committee on Bioethics, *Legislation on Biotechnologies in the Nordic Countries – An overview 2022* (Nordforsk, 2022) accessed online at: https://www.norden.org/en/publication/legislation-biotechnology-nordic-countries-overview-2022.

23    P. Brandão, 'European policies on same-sex relationships, adoption and assisted reproduction', *Journal of Reproduction, Contraception, Obstetrics and Gynecology* 11(8) (2022): 2308–11.

24    International Federation of Fertility Societies' Surveillance (IFFS), 'Global trends in reproductive policy and practice', e58, Autumn 2022: 58.

25    E. Belmonte, M. Álvarez del Vayo, Á. Bernardo, C. Torrecillas, A. Hernández and L. Laursen, 'More than half of European countries prohibit access to assisted reproduction for lesbians and almost a third do so for single women', CIVIO: Medicamentalia, 2 November 2021. Accessed online: https://civio.es/medicamentalia/2021/11/02/ART-EU-access/.

26    N. Cross, 'Exclusion and ignorance: International legal recognition and criminalisation responses to transgender communities in the context of political economy', in H. Panter and A. Dwyer (eds), *Transgender People and Criminal Justice. Critical Criminological Perspectives* (Cham: Palgrave Macmillan, 2023), 53.

27    M. Salama, V. Isachenko, E. Isachenko et al., 'Cross border reproductive care (CBRC): A growing global phenomenon with multidimensional implications (a systematic and critical review)', *Journal of Assisted Reproduction and Genetics* 35(7) (2018): 1277–88.

28    https://www.hfea.gov.uk/treatments/embryo-testing-and-treatments-for-disease/approved-pgt-m-and-ptt-conditions/.

29    https://www.hfea.gov.uk/treatments/embryo-testing-and-treatments-for-disease/pre-implantation-genetic-testing-for-monogenic-disorders-pgt-m-and-pre-implantation-genetic-testing-for-chromosomal-structural-rearrangements-pgt-sr/.

30    https://www.legislation.gov.uk/ukpga/2008/22/section/14.

31    https://www.legislation.gov.uk/ukpga/2008/22/section/14.

32    Paragraph 110 of the original version of the Explanatory Notes to the Human Fertilisation and Embryology Bill as cited in Gerald Porter and Malcolm K. Smith, 'Preventing the selection of "deaf embryos" under the human fertilisation and embryology act 2008: Problematizing disability?' *New Genetics and Society* 32(2) (2013): 172.

33    See for instance: A. V. Middleton, J. V. Hewison and R. V. Mueller, 'Prenatal diagnosis for deafness – what is the potential demand?', *Journal of Genetic Counselling* 10 (2001): 121–31; S. J. Stern, K. S. Arnos, L. Murrelle, K. O. Welch, W. E. Nance and A. Pandya, 'Attitudes of deaf and hard of hearing subjects towards genetic testing and prenatal diagnosis of hearing loss', *Journal of Medical Genetics* 39 (2002): 449–53.

34    S. Segers, G. Pennings and H. Mertes, 'Getting what you desire: the normative significance of genetic relatedness in parent–child relationships', *Medical Health Care and Philosophy* 22 (2019): 487–95; J. McCandless and S. Sheldon, 'Genetically challenged: The determination of legal parenthood in assisted reproduction', in T. Freeman, S. Graham, F. Ebtehaj and M. Richards (eds), *Relatedness in Assisted Reproduction: Families, Origins and Identities* (Cambridge: Cambridge University Press, 2014), 65.

35   M. E. C. Ginoza and R. Isasi, 'Regulating preimplantation genetic testing across the world: A comparison of international policy and ethical perspectives', *Cold Spring Harbour Perspective in Medicine* 10(5) (May 2020): 4.

36   Ginoza and Isasi, 'Regulating preimplantation genetic testing across the world', 6.

37   See, for example, I. S. Markham, 'Ethical and legal issues', *British Medical Bulletin* 54(4) (1998): 1011 and Royal College of Obstetricians and Gynaecologists, 'Supporting women and their partners through prenatal screening for Down's syndrome, Edwards' syndrome and Patau's syndrome', 2020, available online: https://www.rcog.org.uk/guidance/browse-all-guidance/other-guidelines-and-reports/supporting-women-and-their-partners-through-prenatal-screening-for-downs-syndrome-edwards-syndrome-and-pataus-syndrome/.

38   K. M. Jensen et al., 'Low rates of preventive healthcare service utilization among adolescents and adults with down syndrome', *American Journal of Preventive Medicine* 60(1) (2021): 1–12.

39   G. M. Thomas, *Down's Syndrome Screening and Reproductive Politics: Care, Choice, and Disability in the Prenatal Clinic* (Abingdon: Taylor & Francis, 2017), 34

40   NHS, 'Screening for Down's syndrome, Edwards's syndrome and Patau's syndrome', available online: https://www.nhs.uk/pregnancy/your-pregnancy-care/screening-for-downs-edwards-pataus-syndrome/ (accessed 17 October 2023).

41   L. Hui and D. W. Bianchi, 'Fetal fraction and noninvasive prenatal testing: What clinicians need to know', *Prenatal Diagnosis* 40(2) (2020): 155–63.

42   Office for Health Improvement and Disparities, 'NHS screening programmes in England: 2020–2021', updated 16 February 2023, available online, https://www.gov.uk/government/publications/nhs-screening-programmes-annual-report/nhs-screening-programmes-in-england-2020-to-2021 (accessed on 17 October 2023).

43   See, for example, J. K. Morris and A. Springett, *The National Down Syndrome Cytogenetic Register for England and Wales 2013 Annual Report* (Queen Mary University of London, Barts and The London School of Medicine and Dentistry 2014), 1 and A. Wahlberg and T. M. Gammeltoft (eds), *Selective Reproduction in the 21st Century* (Springer International Publishing AG, 2017), 74.

44   National Collaborating Centre for Women's and Children's Health (UK). *Antenatal Care: Routine Care for the Healthy Pregnant Woman* (London: RCOG Press, 2008). March. PMID: 21370514, 174–6; M. L. Constantine, M. Allyse, M. Wall, R. D. Vries and T. H. Rockwood, 'Imperfect informed consent for prenatal screening: Lessons from the quad screen', *Clinical Ethics* 9(1) (2014): 17–27 at p. 23.

45   See M. Brazier, E. Cave and R. Heywood, *Medicine, Patients and the Law*, 7th edition (Manchester: Manchester University Press, 2023) Chapter 6.

46   NHS guidance, 'Screening for Down's syndrome, Edwards' syndrome and Patau's syndrome', available online: https://www.nhs.uk/pregnancy/your-pregnancy-care/screening-for-downs-edwards-pataus-syndrome/.

47	See, for example: National Collaborating Centre for Women's and Children's Health (UK), *Antenatal Care: Routine Care for the Healthy Pregnant Woman*, 174–6; Celine Lewis, Melissa Hill and Lyn S. Chitty, 'Offering non-invasive prenatal testing as part of routine clinical service. Can high levels of informed choice be maintained?' *Prenatal Diagnosis* 37(11) (2017): 1130–7; R. V. van Schendel, G. C. Page-Christiaens, L. Beulen et al., 'Trial by Dutch laboratories for evaluation of non-invasive prenatal testing. Part II women's perspectives', *Prenatal Diagnosis* 36 (2016): 1091–8. PMID: 21370514.

48	M. L. Constantine, M. Allyse, M. Wall, R. D. Vries and T. H. Rockwood, 'Imperfect informed consent for prenatal screening: Lessons from the quad screen', *Clinical Ethics* 9(1) (2014): 17–27 at p. 23.

49	National Collaborating Centre for Women's and Children's Health (UK), *Antenatal Care: Routine Care for the Healthy Pregnant Woman* (London: RCOG Press, 2008), 174–6.

50	H. Statham and W. Solomou, 'Antenatal screening for Down's syndrome', *The Lancet* 352(9143) (1998): 1862.

51	Nancy Press and C. H. Browner 'Why women say yes to prenatal diagnosis', *Social Science Medicine* 45(7) (1997): 987–8.

52	C. Williams, P. Alderson and B. Farsides, 'Too many choices? Hospital and community staff reflect on the future of prenatal screening', *Social Science & Medicine* 55 (2002): 747.

53	C. Lewis, C. Silcock and L. S. Chitty, Non-invasive prenatal testing for Down's syndrome: Pregnant women's views and likely uptake', *Public Health Genomics* 16(5) (2013): 226.

54	V. Ravitsky, S. Birko, J. Le Clerc-Blain, H. Haidar, A. O. Affdal, M. È. Lemoine and A. M. Laberge, 'Noninvasive prenatal testing: Views of Canadian Pregnant women and their partners regarding pressure and societal concerns', *AJOB Empirical Bioethics* 12(1) (2020): 57.

55	See for instance: https://phescreening.blog.gov.uk/2019/06/04/language-matt ers-especially-if-youre-a-health-professional-talking-to-parents-to-be/.

56	Anon., 'Steep rise in Down's pregnancies', BBC News website, 27 October 2009 http://news.bbc.co.uk/1/hi/health/8327228.stm (accessed 20 March 2023).

57	P. Nakou, 'Is routine prenatal screening and testing fundamentally incompatible with a commitment to reproductive choice? Learning from the historical context', *Medicine, Health Care and Philosophy* 24(1) (2021): 73–83; V. Ravitsky, 'The shifting landscape of prenatal testing: Between reproductive autonomy and public health', *Hastings Center Report* 47 (2017): S36.

58	V. Ravitsky, 'The shifting landscape of prenatal testing: Between reproductive autonomy and public health', *Hastings Center Report* 47 (2017): S34.

59	See for instance Z. Stein, M. Susser and A. V. Guterman, 'Screening programme for prevention of Down's Syndrome', *The Lancet (British edition)* 1(7798) (1973): 305–10.

60	V. Seavilleklein, 'Challenging the Rhetoric of choice in prenatal screening', *Bioethics* 23(1) (2009): 71.

61  *The Global Prenatal, Maternal and Newborn Screening Diagnostic Test Market Will Be Worth over $0.5 Billion by 2021* (Coventry: Normans Media Ltd, 2016).

62  G. Stapleton, 'Qualifying choice: Ethical reflection on the scope of prenatal screening', *Medicine, Health Care and Philosophy* 20(2) (2017): 195–205.

63  Nakou, 'Is routine prenatal screening and testing fundamentally incompatible with a commitment to reproductive choice', 73–83.

64  K. L. Lawson, K. Carlson and J. M. Shynkaruk, 'The portrayal of Down syndrome in prenatal screening information pamphlets', *Journal of Obstetrics and Gynaecology Canada* 34(8) (August 2012): 760–8; S. Hall, L. Chitty, E. Dormandy et al., 'Undergoing prenatal screening for Down's syndrome: Presentation of choice and information in Europe and Asia', *European Journal of Human Genetics* 15 (2007): 563–9.

65  N. Press and C. Browner, 'Collective silences, collective fictions: How prenatal diagnostic testing became part of routine prenatal care', in K. H. Rothenberg and E. J. Thomson (eds), *Women and Prenatal Testing: Facing the Challenges of Genetic Technology* (Columbus: Ohio State University Press, 1994), 202–3.

66  Nuffield Council on Bioethics, *Non-Invasive Prenatal Testing: Ethical Issues* (Nuffield Council on Bioethics, 2017), 133.

67  R. Bennett, 'Routine antenatal HIV testing and informed consent: An unworkable marriage?' *Journal of Medical Ethics* 33(8) (2007): 446–8.

68  R. Bennett, 'Antenatal genetic testing and the right to remain in ignorance', *Theoretical Medicine and Bioethics* 22(5) (2001): 469.

69  V. Okur and W. K.Chung, 'Practicing prenatal medicine in a genomic future: How the Practice of pediatrics may (or may not) change with the introduction of widespread prenatal sequencing', in M. A. Allyse and M. Michie (eds), *Born Well: Prenatal Genetics and the Future of Having Children* (Cham: Springer, 2022), 15–29.

70  See, for example, Z. A. Stein and M. Susser, 'The preventability of Down's syndrome', *HSMHA Health Reports* 86 (1971): 650–8.

71  See, for example, K. Krell, K. Haugen, A. Torres and S. L. Santoro, 'Description of daily living skills and independence: A cohort from a multidisciplinary Down Syndrome clinic', *Brain Science* 11(8) (30 July 2021): 1012; J. Docherty and K. Reid, ' "What's the next stage?" Mothers of young adults with Down Syndrome explore the path to independence: A qualitative investigation', *Journal of Applied Research in Intellectual Disabilities* 22(5) (2009): 465.

72  P. Alderson, 'Down's Syndrome: Cost, quality and value of life', *Social Science & Medicine* 53(5) (2001): 627.

73  See, for example, Nancy Press and C. H. Browner, 'Why women say yes to prenatal diagnosis', *Social Science Medicine* 45(7) (1997): 988.

74  Bennett, 'Antenatal genetic testing and the right to remain in ignorance', 461–71.

75  M. Brazier, A. Campbell and S. Golombok, *Surrogacy: Review for Health Ministers of Current Arrangements for Payments and Regulation – Report of the Review Team.* Department of Health and Social Care, 1998, 37–8. http://web

archive.nationalarchives.gov.uk/+/www.dh.gov.uk/en/Publicationsandstatist
ics/Publications/PublicationsLegislation/DH_4009697.

76  See, for example: J. M. Appel, 'Deaf parents want deaf baby: Bioethicist weighs in – you voted now see the results and an expert's discussion', *Medpage Today*, 11 October 2019, accessed online at: https://www.medpagetoday.com/publich ealthpolicy/ethics/82690.

77  D. Teather, 'Lesbian couple have deaf baby by choice', *The Guardian*, 8 April 2002, accessed online at: https://www.theguardian.com/world/2002/apr/08/ davidteather.

78  M. Pyeatt, 'Deaf lesbians criticized for efforts to create deaf child', *Baptist Press*, 3 April 2002, accessed online at: https://www.baptistpress.com/resource-libr ary/news/deaf-lesbians-criticized-for-efforts-to-create-deaf-child/.

79  J. Langton, 'Lesbians: We made our baby deaf on purpose', *The Evening Standard*, 7 April 2002, accessed online at: https://www.standard.co.uk/hp/ front/lesbians-we-made-our-baby-deaf-on-purpose-6311027.html.

80  J. Harris, 'Is there a coherent social conception of disability?' *Journal of Medical Ethics* 26 (2000): 95–100; J. Savulescu, 'Procreative beneficence: Why we should select the best children', *Bioethics* 15 (2001): 413–26.

81  R. Bennett and J. Harris, 'Are there lives not worth living? When is it morally wrong to reproduce?', in D. Dickenson (ed.), *Ethical Issues in Maternal-Fetal Medicine* (Cambridge: Cambridge University Press, 2002), 321–34 at p. 325.

82  J. Savulescu and G. Kahane, 'The moral obligation to create children with the best chance of life', *Bioethics* 23(5) (2009): 277.

83  J. Glover, 'Future people, disability and screening', in John Harris (ed.), *Bioethics* (Oxford: Oxford University Press, 2001), 438.

84  See for example: E. Lee, J. Macvarish and S. Sheldon, 'Assessing child welfare under the Human Fertilisation and Embryology Act 2008: A case study in medicalisation?' *Sociology of Health & Illness* 36(4) (2014): 504; J. E. Stern, C. P. Cramer et al., 'Determining access to assisted reproductive technology: Reactions of clinic directors to ethically complex case scenarios', *Human Reproduction* 18(6): 1343–52.

85  K. Ehrich, C. Williams, R. Scott, J. Sandall and B. Farsides, 'Social welfare, genetic welfare? Boundary-work in the IVF/PGD clinic', *Social Science & Medicine* 63 (2006): 1217.

86  R. Gray, 'Couples could win right to select deaf baby in embryo Bill change', *The Sunday Telegraph*, 13 April 2008.

87  'Deaf parents could choose to have deaf children', *The Independent*, 21 September 2000.

88  See, for example, E. Dormandy and T. M. Marteau, 'Uptake of a prenatal screening test: The role of healthcare professionals' attitudes towards the test', *Obstetrical & Gynecological Survey* 60(5) (2005): 286–7.

89  C. D. Roberts, L. M. Stough and L. H. Parrish, 'The role of genetic counseling in the elective termination of pregnancies involving fetuses with disabilities', *The Journal of Special Education* 36(1) (2002): 49–51.

## CHAPTER 2


1   The HFEA code of Practice, 9th edition. London: HFEA Section 8.3 https://por tal.hfea.gov.uk/knowledge-base/read-the-code-of-practice/.

2   L. M. Osbeck and B. S. Held, 'Introduction', in L. M. Osbeck and B. S. Held (eds), *Rational Intuition: Philosophical Roots, Scientific Investigations* (Cambridge: Cambridge University Press, 2014), 11–12.

3   J. J. Thompson, 'A defense of abortion', *Philosophy and Public Affairs* 1 (1971): 47–66. Thomson's thought experiment asks us to imagine that we wake up and find that, without our consent, we are a life support machine for unconscious violinist. The violinist needs to use our kidneys for nine months or will die. Thompson asks whether it is morally acceptable to disconnect the violinist and allow them to die. She argues that, despite the violinist being a person with an interest in continuing their life, as we have not consented to this use of our body, we are not morally obliged to provide this life support. The point she makes with this thought experiment when it comes to abortion is that even if we put to one side the difficult issue of the moral status of the fetus and assume that it can be considered to have the same rights as the violinist, this does not give mean that a pregnant person is necessarily morally obligated to provide life support for this fetus, particularly if attempts were made to avoid pregnancy.

4   L. R. Kass, 'The wisdom of repugnance', *The New Republic* 216(22) (1997): 20.

5   J. Harris, 'What it's like to be good', *Cambridge Quarterly of Healthcare Ethics* 21 (2012): 294.

6   Harris, 'What it's like to be good', 296.

7   T. D. Wilson, *Strangers to Ourselves: Discovering the Adaptive Unconscious* (Cambridge, MA: Harvard University Press, 2002), 16.

8   Wilson, *Strangers to Ourselves*, 6.

9   Wilson, *Strangers to Ourselves*, 8.

10  H. Oberai and I. Mehrotra Anand, 'Unconscious bias: Thinking without thinking', *Human Resource Management International Digest* 26(6) (2018): 14–17.

11  K. Nalty, 'Strategies for confronting unconscious bias', *The Colorado Lawyer* 45(5) (May 2016): 45.

12  Nalty, 'Strategies for confronting unconscious bias', 46.

13  Oberai and Mehrotra Anand, 'Unconscious bias', 15.

14  Oberai and Mehrotra Anand, 'Unconscious bias', 15.

15  Oberai and Mehrotra Anand, 'Unconscious bias', 14.

16  See, for instance, Implicit Association Test (IAT) at Project Implicit https:// implicit.harvard.edu/implicit/.

17  J. F. Irwin and D. L. Real, 'Unconscious influences on judicial decision-making: The illusion of objectivity', *McGeorge Law Review* 42(1) (2010): 3.

18  See, for example, London School of economics and Political Science, *Avoiding Unconscious Bias*, 5 available online: https://info.lse.ac.uk/staff/divisions/Human-Resources/Assets/Documents/Recruitment-Toolkit/Candidate-Select ion/3.1-Avoiding-Unconscious-Bias.pdf.

19 British Library, *A Timeline of LGBTQ communities in the UK*. Available online: https://www.bl.uk/lgbtq-histories/lgbtq-timeline.

20 G. M. Herek, 'The psychology of sexual prejudice', *Current Directions in Psychological Science: A Journal of the American Psychological Society* 9(1) (2000): 20.

21 M. Atkinson, 'Homosexual law reform', *University of Tasmania Law Review* 11(2) (1992): 206.

## CHAPTER 3

1 Genesis 1:28; Quran, Surah Al-Furqan 25:74, al-Saffat 37:100, Maryam 19:3-7, https://quran.com/25; Hinduism: Ritual to Conceive a Child (Garbhadhana Samskara) which is a prayer to gods, available online: https://www.indastro.com/astrology-articles/ritual-to-conceive-a-child-garbhadhana-samskara-.html#:~:text=Sankalpa%20(Resolve)%3A%20The%20man,in%20her%20womb%20as%20well.

2 See Ipsos, Global happiness up six points since last year: 73% now say they are happy, 20 March 2023, available online: https://www.ipsos.com/en-uk/ipsos-global-happiness-index-2023 (accessed on 26 October 2023); J. F. Helliwell, R. Layard, J. D. Sachs, J.-E. De Neve, L. B. Aknin and S. Wang (eds), *World Happiness Report 2023* (New York: Sustainable Development Solutions Network, 2023).

3 P. Schönegger, 'What's up with anti-natalists? An observational study on the relationship between dark triad personality traits and anti-natalist views', *Philosophical Psychology* 35(1) (2022): 66–94, DOI: 10.1080/09515089.2021.1946026; S. Woolfe, 'On Antinatalism and Depression' Epoché (2020), available online: https://epochemagazine.org/30/on-antinatalism-and-depression/.

4 H. Sherwood, 'Choosing pets over babies is "selfish and diminishes us", says Pope', *The Guardian*, 5 January 2022; R. Gillespie, 'Voluntary childlessness in the United Kingdom', *Reproductive Health Matters* 7(13) (1999): 44.

5 See, for example, J. Harris, *Wonderwoman and Superman: The Ethics of Human Biotechnology* (Oxford: Oxford University Press, 1992), 96; D. Parfit, *Reasons and Persons* (Oxford: Clarendon Press, 1987), 358; J. Feinberg, 'Wrongful life and the counterfactual element in harming', *Social Philosophy and Policy* 4(1) (1986): 160; A. Buchanan, 'Choosing who will be disabled: Genetic intervention and the morality of inclusion', *Social Philosophy and Policy* 13(2) (1996): 28. doi: 10.1017/S0265052500003447.

6 See, for example, Harris, *Wonderwoman and Superman*, 96; Parfit, *Reasons and Persons*, 358.

7 J. Glover, *Causing Death and Saving Lives* (London: Penguin, 1988), 52.

8    R. Bennett and J. Harris, 'Are there lives not worth living? When is it morally wrong to reproduce', in Donna Dickenson (ed.), *Ethical Issues in Maternal-Fetal Medicine* (Cambridge: Cambridge University Press, 2002), 323.

9    E. Silvestri, 'Lebensunwertes leben: Roots and memory of Aktion T4', *Conatus* 4(2) (2019): 65–82.

10    J. Daryl Charles, 'Lebensunwertes leben: The devolution of personhood in the weimar and pre-weimar era', *Ethics Medicine* 21(1) (Spring 2005): 44.

11    L. Pattison, 'The Cure's Robert Smith: "I'm uncomfortable with politicised musicians"', *The Guardian*, 10 September 2011. Accessed online at: https://www.theguardian.com/music/2011/sep/10/robert-smith-the-cure-bestival.

12    D. Benatar, *Better Never to Have Been: The Harm of Coming into Existence* (Oxford: Oxford University Press, 2008).

13    Benatar, *Better Never to Have Been*, 1.

14    M. Häyry, 'A rational cure for prereproductive stress syndrome', *Journal of Medical Ethics* 30 (2004): 377.

15    Benatar, *Better Never to Have Been*, 1.

16    Benatar, *Better Never to Have Been*, 92.

17    Benatar, *Better Never to Have Been*, 23.

18    Parfit, *Reasons and Persons*, 391; Benatar, *Better Never to Have Been*, 30–40.

19    Benatar, *Better Never to Have Been*, 92.

20    Benatar, *Better Never to Have Been*, 224.

21    S. Shead, 'Climate change is making people think twice about having children'. CNBC Sustainable Future 2021 12 August. Accessed online at: https //www.cnbc.com/2021/08/12/climate-change-is-making-people-think-twice-about-having-children.html.

22    S. Aksoy, 'Response to: A rational cure for pre-reproductive stress syndrome', *Journal of Medical Ethics* 30(4) (2004): 383.

23    J. Harris, 'Is there a coherent social conception of disability?' *Journal of Medical Ethics* 26 (2000): 97.

24    J. Herring, *Medical Law and Ethics*, 9th edition (Oxford: Oxford University Press, 2022), 368.

25    M. Strasser, 'Wrongful life, wrongful birth, wrongful death, and the right to refuse treatment: Can reasonable jurisdictions recognize all but one?' *Missouri Law Review* 64 (1999): 63.

26    D. Heyd, 'The intractability of the nonidentity problem', in M. A. Roberts and D. T. Wasserman (eds), *Harming Future Persons* (Dordrecht: Springer Netherlands, 2009), 15.

27    Heyd, 'The intractability of the nonidentity problem', 15.

28    Heyd, 'The intractability of the nonidentity problem', 15.

29    Of course, there will be existing individuals who regret the fact that they have not been able to have children. But this regret about lacking the experience of parenting does not mean that anything bad has been done to another entity by not conceiving and bringing to birth possible children.

## Notes

30   J. F. Helliwell, R. Layard, J. D. Sachs, J.-E. De Neve, L. B. Aknin and S. Wang (eds), *World Happiness Report 2023* (New York: Sustainable Development Solutions Network, 2023), 21.

31   Benatar, *Better Never to Have Been*, Chapter 3.

32   Although having a disability does not necessarily mean that parenting will be more challenging.


## CHAPTER 4

1   D. Parfit, *Reasons and Persons* (Oxford: Clarendon Press, 1987), 351.

2   See, for example, S. Glenn and C. Cunningham, 'Evaluation of self by young people with Down Syndrome', *International Journal of Disability, Development and Education* 48(2) (2001): 173;  G. L. Albrecht and P. J. Devlieger, 'The disability paradox: High quality of life against all odds', *Social Science & Medicine (1982)* 48(8) (1999): 977–88.

3   J. Savulescu and G. Kahane, 'The moral obligation to create children with the best chance of life', *Bioethics* 23(5) (2009): 277.

4   Parfit, *Reasons and Persons*, 451.

5   Parfit, *Reasons and Persons*, 361.

6   R. Bennett, 'The fallacy of the principle of procreative beneficence', *Bioethics* 23(5) (June 2009): 270–1.

7   J. Savulescu and G. Kahane, 'Understanding procreative beneficence', in Leslie Francis (ed.), *The Oxford Handbook of Reproductive Ethics* (Oxford: Oxford University Press, 2017), 23. See also J. Harris, 'Is there a coherent social conception of disability?', *Journal of Medical Ethics* 26 (2000): 96.

8   R. Sparrow, 'A not-so-new EUGENICS: Harris and Savulescu on human enhancement', *The Hastings Center Report* 41(1) (January–February 2011): 32–42; R. Bennett, 'When intuition is not enough. Why the principle of procreative beneficence must work much harder to justify its eugenic vision', *Bioethics* 28(9) (2014): 447–55.

9   Parfit, *Reasons and Persons*.

10   J. Harris, *The Value of Life: An Introduction to Medical Ethics* (London: Routledge & Kegan Paul, 1985), 146–9.

11   R. Bennett and J. Harris, 'Are there lives not worth living? When is it morally wrong to reproduce?', in D. Dickenson (ed.), *Ethical Issues in Maternal-Fetal Medicine* (Cambridge: Cambridge University Press, 2002), 321–34, p. 325.

12   J. Savulescu, 'Procreative beneficence: Why we should select the best children', *Bioethics* 15 (2001): 413–26.

13   Savulescu, 'Procreative beneficence', 413–26, 413.

14   R. Bennett and J. Harris, 'Are there lives not worth living?', 325; M. Parker, 'The best possible child', *Journal of Medical Ethics* 33 (2007): 279–83.

15   Parfit, *Reasons and Persons*, 358.

16   Parfit, *Reasons and Persons*, 358–61.

17   Parfit, *Reasons and Persons*, 387.

18   J. Harris, *Clones, Genes and Immortality* (Oxford: Oxford University Press, 1998), 117. Interestingly Harris does argue that individuals with any impairment however minor have been harmed by this condition but that this harm does not mean that we have done anything wrong to that person in terms of concerns for their overall welfare. He argues that they are harmed but not wronged see Bennett and Harris, 'Are there lives not worth living?', 321–34, p. 325.

19   Bennett and Harris, 'Are there lives not worth living?', 321–34, p. 325.

20   Savulescu, 'Procreative beneficence', 423.

21   Savulescu, 'Procreative beneficence', 418.

22   Parfit, *Reasons and Persons*, 351.

23   Parfit, *Reasons and Persons*, 388.

24   Parker, 'The best possible child', 279–83.

25   Sparrow, 'A not-so-new EUGENICS: Harris and Savulescu on human enhancement', 32–42; Bennett, 'When intuition is not enough', 447–55.

26   Savulescu, 'Procreative beneficence', 424.

27   Savulescu, 'Procreative beneficence', 424.

28   Harris, 'Is there a coherent social conception of disability?', 95.

29   Harris, 'Is there a coherent social conception of disability?', 96.

30   J. Harris, 'One principle and three fallacies of disability studies', *Journal of Medical Ethics* 27(6) (December 2001): 383.

31   Bennett and Harris, 'Are there lives not worth living?', 352.

32   D. Benetar, *Better Never to Have Been: The Harm of Coming into Existence* (Oxford: Oxford University Press, 2006), 64–9.

33   M. Häyry, 'If you must make babies, then at least make the best babies you can?' *Human Fertility* 7(2) (2004): 105–12.

34   H. Lane, 'Do deaf people have a disability?' *Sign Language Studies* 2(4) (2002): 364. *JSTOR*, http://www.jstor.org/stable/26204820. Accessed 30 October 2023.

35   Bennett and Harris, 'Are there lives not worth living?', 325.

36   Harris, 'Is there a coherent social conception of disability?', 95–100, 97.

37   J. Harris, 'Rights and reproductive choice', in J. Harris and S. Holm (eds), *The Future of Human Reproduction* (Oxford: Oxford University Press, 1998), 31.

38   Harris, 'Is there a coherent social conception of disability?', 97.

39   Parfit, *Reasons and Persons*, 351.

40   F. Janssen-Lauret, 'Anti-essentialism, modal relativity, and alternative material-origin counterfactuals', *Synthese (Dordrecht)* 199(3–4) (2021): 8379.

41   Janssen-Lauret, 'Anti-essentialism, modal relativity, and alternative material-origin counterfactuals', 8379.

42   D. McClean, 'A moral requirement for energy policies', in D. McClean and P. Brown (eds), *Energy and the Future* (Totowa, NJ: Rowman & Littlefield, 1983), 196 cited in C. Wolf, 'Do future persons presently have alternative possible

identities?', in M. A. Roberts and D. T. Wasserman (eds), *Harming Future Persons* (New York: Springer, 2009), 106.

43   J. David Velleman, 'Persons in prospect', *Philosophy & Public Affairs* 36 (2008): 275.

44   D. S. Davis, 'Genetic dilemmas and the child's right to an open future', *The Hastings Center Report* 27(2) (1997): 11.

45   B. Steinbock, 'Wrongful life and procreative decision', in Melinda A. Roberts and David T. Wasserman (eds), *Harming Future Persons: Ethics, Genetics and the Nonidentity Problem* (New York: Springer, 2009), 163.

46   D. Archard, 'Wrongful life', *Philosophy* 79(309) (2004): 406.

47   See, for example, Y. Huang, H. Liu and M. Masum, 'Adverse childhood experiences and physical and mental health of adults: Assessing the mediating role of cumulative life course poverty', *American Journal of Health Promotion* 35(5) (2021): 637–47.

48   H. Statham and W. Solomou, 'Antenatal screening for Down's syndrome', *The Lancet* 352(9143) (1998): 1862; V. Ravitsky, S. Birko, J. Le Clerc-Blain, H. Haidar, A. O. Affdal, M. È. Lemoine, C. Dupras and A. M. Laberge, 'Noninvasive prenatal testing: Views of Canadian pregnant women and their partners regarding pressure and societal concerns', *AJOB Empirical Bioethics* 12(1) (2020): 57.

## CHAPTER 5

1   J. Savulescu, 'Procreative beneficence: Why we should select the best children', *Bioethics* 15 (2001): 417.

2   J. Harris, 'One principle and three fallacies of disability studies', *Journal of Medical Ethics* 27(6) (2001): 385.

3   Savulescu, 'Procreative beneficence', 416.

4   Harris, 'One principle and three fallacies of disability studies', 385.

5   J. Harris, *Wonderwoman and Superman* (Oxford: Oxford University Press 1992), 72–3.

6   Savulescu, 'Procreative beneficence', 416.

7   For more detail on this issue see R. Bennett, 'When intuition is not enough. Why the principle of procreative beneficence must work much harder to justify its eugenic vision', *Bioethics* 28(9) (2014): 447–55.

8   R. Bennett and J. Harris, 'Are there lives not worth living? When is it morally wrong to reproduce?', in D. Dickenson (ed.), *Ethical Issues in Maternal-Fetal Medicine* (Cambridge: Cambridge University Press, 2002), 352.

9   Savulescu, Procreative beneficence', 418.

10   J. Savulescu and G. Kahane, 'The moral obligation to create children with the best chance of the best life', *Bioethics* 23(5) (2009): 277.

11   J. Moule, 'Understanding unconscious bias and unintentional racism', *Phi Delta Kappan* 90(5) (2009): 322.

12    Moule, 'Understanding unconscious bias and unintentional racism', 325.

13    B. Applebaum, 'Good, liberal intentions are not enough: Racism, intentions, and moral responsibility', *Journal of Moral Education* 26 (December 1997): 409.

14    C. Friedman, 'Defining disability: Understandings of and attitudes towards ableism and disability', *Disability Studies Quarterly* 37(1) (2017) no page number available for online version.

15    Friedman, 'Defining disability', no page number available for online version.

16    C. Friedman, 'Mapping ableism: A two-dimensional model of explicit and implicit disability attitudes', *Canadian Journal of Disability Studies* 8(3) (May 2019): 95.

17    J. F. Dovidio, L. Pagotto and M. R. Hebl, 'Implicit attitudes and discrimination against people with physical disabilities', in R. Wiener and S. Wilborn (eds), *Disability and Aging Discrimination: Perspectives in Law and Psychology* (New York: Springer, 2011), 171.

18    Dovidio, Pagotto and Hebl, 'Implicit attitudes and discrimination against people with physical disabilities', 174 and 167.

19    Dovidio, Pagotto and Hebl, 'Implicit attitudes and discrimination against people with physical disabilities', 170.

20    Dovidio, Pagotto and Hebl, 'Implicit attitudes and discrimination against people with physical disabilities', 161.

21    Dovidio, Pagotto and Hebl, 'Implicit attitudes and discrimination against people with physical disabilities', 160.

22    Friedman, 'Defining disability', no page number available for online version.

23    Dovidio, Pagotto and Hebl, 'Implicit attitudes and discrimination against people with physical disabilities', 176.

24    D. S. Davis, *Genetic Dilemmas: Reproductive Technology, Parental Choices, and Children's futures*, 2nd edition (Oxford: Oxford University Press, 2010), 19–20.

25    Savulescu and Kahane, 'The moral obligation to create children with the best chance of life', 277.

26    Harris, *Wonderwoman and Superman*, 72–3.

27    G. L. Albrecht and P. J. Devlieger, 'The disability paradox: High quality of life against all odds', *Social Science & Medicine* 48 (1999): 977–88.

28    T. Shakespeare, 'A point of view: Happiness and disability', *BBC News* 1 June 2014, available online: https://www.bbc.co.uk/news/magazine-27554754.

29    See, for example, J.-H. Lin, Y.-H. Ju, S.-J. Lee, C.-H. Lee, H.-Y. Wang, Y.-L. Teng and S. K. Lo, 'Do physical disabilities affect self-perceived quality of life in adolescents?' *Disability and Rehabilitation* 31(3) (2009): 181–8; J. O'Hara, A. P. Martin, D. Nugent, M. Witkop, T. W. Buckner, M. W. Skinner, B. O'Mahony, B. Mulhern, G. Morgan, N. Li and E. K. Sawyer, 'Evidence of a disability paradox in patient-reported outcomes in haemophilia', *Haemophilia* 27 (2021): 245–52.

30    C. E. Schwartz and M. A. Sprangers, 'Methodological approaches for assessing response shift in longitudinal health-related quality-of-life research', *Social Science & Medicine* 48 (1999): 1531–48.

31    P. Ubel, George Loewenstein, Norbert Schwarz and Dylan Smith, 'Misimagining the unimaginable: The disability paradox and health care decision making', *Health Psychology* 24(4) (2005): S57–S62: S62.

32    Shakespeare, 'A point of view'.

33    Of course, those who take a different view of the moral status of embryo and fetus may argue that comparisons can be made here as they may view the person as already existing at this early stage. See Chapter 1 (pp. 15–18) for a discussion on this and the concept of future persons.

34    J. Harris, 'Is there a coherent social conception of disability?' *Journal of Medical Ethics* 26 (2000): 95–100, p. 97.

35    Bennett and Harris, 'Are there lives not worth living?', 325.

36    Bennett and Harris, 'Are there lives not worth living?', 325.

37    Although, as we saw in Chapter 3, trying to make a convincing argument that anyone has benefitted from being brought into existence is logically impossible.

38    Harris, 'Is there a coherent social conception of disability?', 95–100, p. 97.

39    See, for example, M. Jay, 'Deaf culture values: Deafness', 15th February 2021, *Start ASL*, accessed online: https://www.startasl.com/deaf-culture-values-deafness/#:~:text=It%20is%20true%20that%20many,identity%2C%20experie nce%2C%20and%20advocacy.

40    Although I still argue we should be careful about moving straight to regulation that involves coercion in these cases see R. Bennett, 'Antenatal genetic testing and the right to remain in ignorance', *Theoretical Medicine and Bioethics* 22 (2001): 461–71.

41    R. Bennett, 'The fallacy of the principle of procreative beneficence', *Bioethics* 23(5) (June 2009): 268–70.

# CHAPTER 6

1    See, for example, B. Steinbock, 'Wrongful life and procreative decision', in M. A. Roberts and D. T. Wasserman (eds), *Harming Future Persons: Ethics, Genetics and the Nonidentity Problem* (New York: Springer, 2009), 163; D. Archard, 'Wrongful life', *Philosophy* 79(309): 406; J. David Velleman, 'Persons in prospect', *Philosophy & Public Affairs* 36 (2008): 275.

2    Steinbock, 'Wrongful life and procreative decision', 155.

3    R. Bennett and J. Harris, 'Are there lives not worth living? When is it morally wrong to reproduce?' in D. Dickenson (ed.), *Ethical Issues in Maternal-Fetal Medicine* (Cambridge: Cambridge University Press, 2002), 321–34, p. 325.

4    M. Häyry, 'A rational cure for prereproductive stress syndrome', *Journal of Medical Ethics* 30 (2004): 377.

5    D. Benatar, *Better Never to Have Been: The Harm of Coming into Existence* (Oxford: Oxford University Press, 2006), 1.

6    Häyry, 'A rational cure for prereproductive stress syndrome', 378.

7  J. Savulescu, 'Procreative beneficence: Why we should select the best children', *Bioethics* 15 (2001): 424.

8  J. Harris, 'Is there a coherent social conception of disability?', *Journal of Medical Ethics* 26 (2000): 95.

9  N. Agar, *Liberal Eugenics: In Defence of Human Enhancement* (Malden, MA: Blackwell, 2005), p. 5.

10  C. MacKellar and C. Bechtel (eds), *The Ethics of the New Eugenics*, 1st edition (New York: Berghahn Books, 2014), 21. *JSTOR*, available online: http://www.jstor.org/stable/j.ctt9qcw9j.7 (accessed 16 October 2023).

11  Harris, 'Is there a coherent social conception of disability?', 96; Savulescu, 'Procreative beneficence', 425.

## CHAPTER 7

1  Human Fertilisation and Embryology Act 1990 s.13(5).

2  Human Fertilisation and Embryology Authority (1991) Code of Practice: Explanation. London: HFEA, 3.15 as cited by Eric Blyth, 'The United Kingdom's human fertilisation and embryology act 1990 and the welfare of the child: A critique', *The International Journal of Children's Rights* 3(3–4) (1995): 435.

3  Blyth, 'The United Kingdom's human fertilisation and embryology act 1990 and the welfare of the child', 422.

4  Lord Mackay *HL Deb* Vol. 516 col. 1098, 1990 (6 March), available online: https://api.parliament.uk/historic-hansard/lords/1990/mar/06/human-fertil isation-and-embryology-bill#column_1098 (accessed on 12 August 2023).

5  Human Fertilisation and Embryology Authority (HFEA), *Tomorrow's Children: A Consultation on Guidance to Licensed Fertility Clinics on Taking in Account the Welfare of Children to be Born of Assisted Conception Treatment* (London: HFEA, 2005).

6  Human Fertilisation and Embryology Authority (HFEA), *Tomorrow's Children*, 1.

7  E. Blyth and C. Cameron, 'The welfare of the child: An emerging issue in the regulation of assisted conception', *Human Reproduction* 13(9) (1998): 2341.

8  S. L. de Lacey, K. Peterson and J. McMillan, 'Child interests in assisted reproductive technology: How is the welfare principle applied in practice?' *Human Reproduction* 30(3) (2015): 617; see also A. D. Gurmankin, A. L. Caplan and A. M. Braverman, 'Screening practices and beliefs of assisted reproductive technology programs', *Fertility and Sterility* 83(1) (2005): 61–7.

9  HFEA (2007) Code of Practice, 7th edition. London: HFEA.

10  The HFEA code of Practice, 9th edition. London: HFEA Section 8.15. Available online: https://portal.hfea.gov.uk/knowledge-base/read-the-code-of-practice/.

11  HFEA (2021) Code of Practice, 9th edition. London: HFEA. Available online: https://portal.hfea.gov.uk/knowledge-base/read-the-code-of-practice/.

12    The HFEA code of Practice, 9th edition. Section 8.14. Available online: https://
      portal.hfea.gov.uk/knowledge-base/read-the-code-of-practice/.

13    H. Priddle, 'How well are lesbians treated in UK fertility clinics?', *Human
      Fertility* 18(3) (2015): 195.

14    F. MacCallum and S. Golombok, 'Children raised in fatherless families from
      infancy: A follow-up of children of lesbian and single heterosexual mothers
      at early adolescence', *Journal of Child Psychology and Psychiatry* 45(8)
      (2004): 1407–19.

15    https://www.legislation.gov.uk/ukpga/2008/22/section/14.

16    https://www.hfea.gov.uk/pgt-m-conditions/.

17    The HFEA code of Practice, 9th edition. Section 10A. Available online: https://
      portal.hfea.gov.uk/knowledge-base/read-the-code-of-practice/.

18    Human Fertilisation and Embryology Authority, 'Pre-implantation genetic
      testing for monogenic disorders (PGT-M) and Pre-implantation genetic
      testing for chromosomal structural re-arrangements (PGT-SR) available
      online: https://www.hfea.gov.uk/treatments/embryo-testing-and-treatme
      nts-for-disease/pre-implantation-genetic-testing-for-monogenic-disord
      ers-pgt-m-and-pre-implantation-genetic-testing-for-chromosomal-structu
      ral-rearrangements-pgt-sr/ (accessed on 16 October 2023).

19    The HFEA code of Practice, 9th edition. Section 10C. Available online: https://
      portal.hfea.gov.uk/knowledge-base/read-the-code-of-practice/.

20    HFEA code of practice 2019, page 109. Available online: https://portal.hfea.
      gov.uk/media/1605/2019-12-03-code-of-practice-december-2019.pdf.

21    The HFEA code of Practice, 9th edition. Section 8.14. Available online: https://
      portal.hfea.gov.uk/knowledge-base/read-the-code-of-practice/.

22    M. Dargis, J. Newman and M. Koenigs, 'Clarifying the link between childhood
      abuse history and psychopathic traits in adult criminal offenders', *Personality
      Disorders: Theory, Research, and Treatment* 7(3) (2016): 221–8.

23    The HFEA code of Practice, 9th edition. Section 8.14. Available online: https://
      portal.hfea.gov.uk/knowledge-base/read-the-code-of-practice/.

24    The HFEA code of Practice, 9th edition. Section 8.14. Available online: https://
      portal.hfea.gov.uk/knowledge-base/read-the-code-of-practice/.

25    The HFEA code of Practice, 9th edition. Section 10.9. Available online: https://
      portal.hfea.gov.uk/knowledge-base/read-the-code-of-practice/.

26    The HFEA code of Practice, 9th edition. Section 10A. Available online: https://
      portal.hfea.gov.uk/knowledge-base/read-the-code-of-practice/.

27    See, for example, M. Jay, 'Deaf culture values: Deafness', 15 February 2021,
      *Start ASL*, accessed online: https://www.startasl.com/deaf-culture-values-
      deafness/#:~:text=It%20is%20true%20that%20many,identity%2C%20experie
      nce%2C%20and%20advocacy.

28    Paragraph 110 of the original version of the Explanatory Notes to the Human
      Fertilisation and Embryology Bill as cited in Gerald Porter and Malcolm
      K. Smith, 'Preventing the selection of "deaf embryos" under the human
      fertilisation and embryology act 2008: Problematizing disability?' *New Genetics
      and Society* 32(2) (2013): 172.

29   National Cancer Institute, 'BRCA Gene Mutations: Cancer Risk and Genetic Testing', accessed online: https://www.cancer.gov/about-cancer/causes-prevent ion/genetics/brca-fact-sheet (accessed on 20 October 2023).

30   The HFEA code of Practice, 9th edition. Section 10A. Available online: https:// portal.hfea.gov.uk/knowledge-base/read-the-code-of-practice/.

31   The HFEA code of Practice, 9th edition. Section 10A. Available online: https:// portal.hfea.gov.uk/knowledge-base/read-the-code-of-practice/.

32   J. Moule, 'Understanding unconscious bias and unintentional racism', *Phi Delta Kappan* 90(5) (2009): 325.

33   E. Lee, J. Macvarish and S. Sheldon, 'Assessing child welfare under the human fertilisation and embryology act 2008: A case study in medicalisation?', *Sociology of Health & Illness* 36(4) (2014): 513.

34   F. MacCallum and S. Golombok, 'Children raised in fatherless families from infancy: A follow-up of children of lesbian and single heterosexual mothers at early adolescence', *Journal of Child Psychology and Psychiatry* 45(8) (2004): 1407–19.

35   HFEA, *Tomorrow's Children* (HFEA, November 2005), 1.

36   L. Frith, 'Process and consensus: Ethical decision-making in the infertility clinic – a qualitative study', *Journal of Medical Ethics* (2009): 663.

37   Frith, 'Process and consensus', 664.

38   Frith, 'Process and consensus', 666.

39   Paragraph 110 of the original version of the Explanatory Notes to the Human Fertilisation and Embryology Bill as cited in G. Porter and M. K Smith, 'Preventing the selection of "deaf embryos" under the human fertilisation and embryology act 2008: problematizing disability?' *New Genetics and Society* 32(2) (2013): 172.

40   E. Jackson, *Medical Law*, 6th edition (Oxford: Oxford University Press, 2022), 52.

41   Lee, Macvarish and Sheldon, 'Assessing child welfare under the human fertilisation and embryology act 2008', 508.

42   https://www.legislation.gov.uk/ukpga/2008/22/section/14.

43   HFEA code of practice 2019, page 109, available online: https://portal.hfea.gov. uk/media/1605/2019-12-03-code-of-practice-december-2019.pdf.

44   https://www.legislation.gov.uk/ukpga/2008/22/section/14.

45   S. Lou, K. Carstensen, O. B. Petersen et al., Termination of pregnancy following a prenatal diagnosis of Down syndrome: A qualitative study of the decision-making process of pregnant couples', *Acta Obstet Gynecol Scand* 97 (2018): 1233–4.

46   R. Bennett, 'Routine antenatal HIV testing and informed consent: an unworkable marriage?', *Journal of Medical Ethics* 33(8) (2007): 446–8.

47   J. Savulescu, 'Procreative beneficence: Why we should select the best children', *Bioethics* 15 (2001): 424.

48   See, for instance, Z. Stein, M. Susser and A. V. Guterman, 'Screening programme for prevention of Down's Syndrome', *The Lancet (British edition)* 1(7798) (1973): 305–10.

49    National Collaborating Centre for Women's and Children's Health (UK), *Antenatal Care: Routine Care for the Healthy Pregnant Woman* (London: RCOG Press, 2008). March. PMID: 21370514, 174–6.

50    M. L. Constantine, M. Allyse, M. Wall, R. D. Vries and T. H. Rockwood, 'Imperfect informed consent for prenatal screening: Lessons from the quad screen', *Clinical Ethics* 9(1) (2014): 17–27 at p. 23.

51    R. Bennett, 'Antenatal genetic testing and the right to remain in ignorance', *Theoretical Medicine and Bioethics* 22(5) (2001): 461–71.

52    See for instance Stein, Susser and Guterman, 'Screening programme for prevention of Down's syndrome', 305–10.

53    S. Sheldon, E. Lee and J. Macvarish, 'Supportive parenting, responsibility and regulation: The welfare assessment under the reformed human fertilisation and embryology act' (1990), *Modern Law Review* 78(3) (2015): 461–92.

54    L. Frith, 'Use or ornament? Clinical ethics committees in infertility units: A qualitative study', *Clinical Ethics* 4(2) (2009): 91–7.

55    S. L. de Lacey, K. Peterson and J. McMillan, 'Child interests in assisted reproductive technology: How is the welfare principle applied in practice?', *Human Reproduction* 30(3) (2015): 620.

56    K. Ehrich, C. Williams, R. Scott, J. Sandall and B. Farsides, 'Social welfare, genetic welfare? Boundary-work in the IVF/PGD clinic', *Social Science & Medicine* 63 (2006): 1219–20.

## CHAPTER 8

1    See Ipsos, Global happiness up six points since last year: 73% now say they are happy, 20 March 2023. Available online: https://www.ipsos.com/en-uk/ipsos-global-happiness-index-2023 (accessed on 26 October 2023); J. F. Helliwell, R. Layard, J. D. Sachs, J.-E. De Neve, L. B. Aknin and S. Wang (eds), *World Happiness Report 2023* (New York: Sustainable Development Solutions Network, 2023).

2    A. Middleton, J. Hewison and R. Mueller, 'Prenatal diagnosis for inherited deafness – what is the potential demand?' *Journal of Genetic Counselling* 10 (2001): 121–31.

3    NHS, 'Screening for Down's syndrome, Edwards's syndrome and Patau's syndrome', available online: https://www.nhs.uk/pregnancy/your-pregnancy-care/screening-for-downs-edwards-pataus-syndrome/ (accessed 17 October 2023).

4    E. J. Lee, J. Macvarish and S. Sheldon, 'Assessing child welfare under the human fertilisation and embryology act: The new law. Summary of findings' (2012): 7. https://blogs.kent.ac.uk/parentingculturestudies/files/2012/06/Summary_Assessing-Child-Welfare-final.pdf.

5    S. S. Baig, M. Strong, E. Rosser, N. V. Taverner, R. Glew, Z. Miedzybrodzka, A. Clarke and D. Craufurd, '22 years of predictive testing for huntington's

disease: The experience of the UK Huntington's prediction consortium', *European Journal of Human Genetics: EJHG* 24(10) (2016): 1396–1402.

6 K. Ehrich, C. Williams, R. Scott, J. Sandall and B. Farsides, 'Social welfare, genetic welfare? Boundary-work in the IVF/PGD clinic', *Social Science & Medicine* 63 (2006): 1218.

7 Ehrich, Williams, Scott, Sandall and Farsides, 'Social welfare, genetic welfare?', 1219.

8 Ehrich, Williams, Scott, Sandall and Farsides, 'Social welfare, genetic welfare?', 1218.

9 E. Lee, J. Macvarish and S. Sheldon, 'Assessing child welfare under the Human Fertilisation and Embryology Act 2008: A case study in medicalisation?', *Sociology of Health & Illness* 36(4) (2014): 513.

10 See, for example, R. Bennett and J. Harris, 'Restoring natural function: Access to infertility treatment using donated gametes', *Human Fertility* 2 (1999): 18–21; E. Jackson, 'Conception and the irrelevance of the welfare principle', *Modern Law Review* 65(2) (March 2002): 176–203: S. Waxman, 'Applying the preconception welfare principle and the harm threshold: Doing more harm than good?', *Medical Law International* 17(3) (2017): 134–57.

11 S. L. de Lacey, K. Peterson and J. McMillan, 'Child interests in assisted reproductive technology: How is the welfare principle applied in practice?', *Human Reproduction* 30(3) (March 2015): 620.

12 See for instance: Middleton, Hewison and Mueller, 'Prenatal diagnosis for deafness – what is the potential demand?', 121–31; S. J. Stern, K. S. Arnos, L. Murrelle, K. O. Welch, W. E. Nance and A. Pandya, 'Attitudes of deaf and hard of hearing subjects towards genetic testing and prenatal diagnosis of hearing loss', *Journal of Medical Genetics* 39 (2002): 449–53.

13 Bennett and Harris, 'Restoring natural function', 20.

# INDEX